Direct Hydroxylation of Methane

Kazunari Yoshizawa
Editor

Direct Hydroxylation of Methane

Interplay Between Theory and Experiment

Springer

Editor
Kazunari Yoshizawa
Institute for Materials Chemistry and
Engineering
Kyushu University
Fukuoka, Japan

ISBN 978-981-15-6985-2 ISBN 978-981-15-6986-9 (eBook)
https://doi.org/10.1007/978-981-15-6986-9

This Springer imprint is published by the registered company Springer Nature Singapore Pte Ltd.
The registered company address is: 152 Beach Road, #21-01/04 Gateway East, Singapore 189721, Singapore

Contents

Orbital Concept for Methane Activation

Kazunari Yoshizawa and Mayuko Miyanishi

Abstract Theoretical thinking on methane C–H bond activation and hydroxylation by soluble and particulate methane monooxygenase (iron and copper enzyme species) and related metal-oxo species such as FeO^+ is developed. The tetrahedral T_d structure of methane can be deformed into a C_{3v} or D_{2d} structure and bound at a coordinatively unsaturated metal-oxo site of a soluble methane monooxygenase model from extended Hückel calculations. Mechanistic aspects about methane hydroxylation by the bare transition-metal oxide ion FeO^+ are analyzed by using density functional theory calculations. An important feature in the reaction is the spin crossover between the high-spin and low-spin potential energy surfaces in particular in the C–H activation process, the energy barrier of which is significantly decreased by the spin inversion. The hydroxylation mechanisms of soluble and particulate methane monooxygenase are considered. These mechanistic insights are reasonably extended to methane activation by metal-exchanged zeolites and IrO_2 (110) surface.

Keywords C–H activation · Density functional theory · Extended Hückel method · Metal oxides · Methane hydroxylation · Methane monooxygenase · Orbital interaction · Zeolites

1 Introduction

Methane and benzene are the most interesting saturated and unsaturated hydrocarbons, respectively. Table 1 lists various properties of methane and benzene. In particular, their C–H bond dissociation energies (BDEs) are extremely large. Benzene's C–H BDE of 110 kcal/mol is slightly larger than methane's C–H BDE of 105 kcal/mol (103 kcal/mol in density functional theory (DFT) calculations). Many important chemical processes starting from benzene are widely used by replacing one of its hydrogen atoms with another functional group. Examples of simple benzene derivatives are phenol, toluene, and aniline that involve OH, CH_3, and NH_2 groups in

K. Yoshizawa (✉) · M. Miyanishi
Institute for Materials Chemistry and Engineering, Kyushu University, Motooka 744, Nishi-ku, Fukuoka 819-0395, Japan
e-mail: kazunari@ms.ifoc.kyushu-u.ac.jp

K. Yoshizawa (ed.), *Direct Hydroxylation of Methane*,
https://doi.org/10.1007/978-981-15-6986-9_1

Table 1 Various properties of methane and benzene

	Methane	Benzene
Melting point (°C)[a]	−182.6	5.5
Boiling point (°C)[a]	−161.6	80.1
Density[a]	0.466 g m^{-3} (at −164 °C)	0.876 g cm^{-3} (at 20 °C)
Critical temperature (°C)[b]	−82.1	289
Critical pressure (atm)[b]	45.8	48.6
First ionization potential (eV)	13.2[c]	9.23[d]
Second ionization potential (eV)	19.4[c]	17.2[e]
First BDE (kcal mol^{-1})[f]	105.1 ± 2	110.9 ± 2
(Calcd. B3LYP/6-311++G**) (kcal mol^{-1})	103.0	110.4
Proton affinity (kcal mol^{-1})[g]	129.9	179.3
pKa[h]	48	43
$\Delta H°_f$[i]	−17.9 kcal mol^{-1} (gas)	11.7 kcal mol^{-1}(liquid) 19.8 kcal mol^{-1}(gas)
C–H bond length (Å)	1.085[j]	1.085[k]
(Calcd. B3LYP/6-311++G**) (Å)	1.091	1.084
C-13 NMR shift (ppm)[l]	−2.3	128.5
J(C–H) (Hz)[l]	125	158
Vibrational frequencies[m,n]	1306 cm^{-1}(degenerate deformation) 1534 cm^{-1} (degenerate deformation) 2917 cm^{-1} (symmetrical stretching) 3019 cm^{-1} (degenerate stretching)	674 cm^{-1} (CH bending) 1038 cm^{-1} (CH bending) 1479 cm^{-1} (ring streching, deforming) 3036 cm^{-1} (CH streching) 3072 cm^{-1} (CH streching) 3091 cm^{-1} (CH streching)

[a]Kaye GWC, Laby TH (1973) Tables of physical and chemical constants. Longman, London
[b]Kobe KA, Lynn RE (1953) Chem Rev 52:117–236
[c]Frost J, McDowell P (1957) Proc Roy Soc A 241:194–207
[d]Arimura M, Yoshikawa Y (1984) Journal of the Mass Spectrometry Society of Japan 32:375–380
[e]Hustruid A, Kusch P, Tate JT (1938) Phys Rev 54:1037–1044
[f]Mamillen DF, Golden DM (1982) Ann Rev Chem 33:493–532
[g]Hunter EPL, Lias SG (1998) J Phys Ref Data 27:413–656
[h]pKa Data Compiled by R. Williams
[i]Atkins PW(1998) Atkins' physical chemistry. Oxword University Press, New York
[j]Kuchitsu K, Bartell LS (1962) J Chem Phys 36:2470–2481
[k]Moe OLE, Stand TG (1985) J molec Struct 128:13–19
[l]Hesse M, Meier H, Zeeh B (1997) Spectroscopic Methods in Organic Chemistry. Georg Thieme Verlag, Stuttgart
[m]Shimanouchi T(1972) Tables of Molecular Vibrational Frequencies. Consolidated Volume 1, NSRDS NBS-39
[n]SDBSWeb: https://sdbs.db.aist.go.jp (National Institute of Advanced Industrial Science and Technology, 2019. 2)

the benzene ring, respectively. The synthesis of aniline from benzene is a simple chemistry experiment for undergraduate students. In contrast, such selective chemical processes for methane under mild conditions have not yet been established even in laboratory.

The selective oxidation of methane [1–5] has attracted increased attention for a long time because of its scientific interest and industrial importance. The direct conversion of methane to methanol (Eq. 1), which is an exothermic process, is catalyzed under physiological conditions by soluble methane monooxygenase (sMMO) [6], which has a diiron active center, and particulate methane monooxygenase (pMMO) [7], which has mono- and dicopper sites. This reaction is also catalyzed by the bare transition-metal-oxide ions such as FeO^+ [8] in the gas-phase and metal-exchanged zeolites such as Fe-ZSM-5 zeolite [9]. The enzymatic reactions of the MMOs use molecular dioxygen as an oxidant, while the latter two reactions are very similar in that nitrous oxide (N_2O) is used as an oxidant. Iron-, nickel-, and copper-oxo (or -oxyl) species are involved in these difficult chemical processes. To develop a man-made catalytic system for this attractive reaction, it is important to reveal the mechanism of the direct hydroxylation processes by these catalytic systems.

$$CH_4 + 1/2\ O_2 \rightarrow CH_3OH\ (\Delta H_0 = -30.7\ \text{kcal/mol}) \quad (1)$$

Quantum chemical calculations and orbital interaction thinking at various levels of theory play a key role in understanding the mechanism of the C–H activation of methane, which is an essential, initial process in the direct methane hydroxylation. However, the mechanism has not been well understood because of its extremely rapid reaction. Figure 1 shows computed C–H BDEs of various alkanes and related hydrocarbons from DFT calculations at the B3LYP/6-311++G** level of theory. In general, the C–H BDEs at the primary (1°), secondary (2°), and tertiary (3°) carbon atoms of alkanes are approximately 97, 94, and 90 kcal/mol, while those at the benzylic positions are less than 85 kcal/mol. In particular, dihydroanthracene and allylbenzne have very weak benzylic C–H bonds of less than 75 kcal/mol. Only methane has a C–H BDE of over 100 kcal/mol among these alkanes.

The C–H BDEs also have a good correlation with computed HOMO–LUMO energy gaps, as shown in Fig. 2. Methane's HOMO–LUMO gap is calculated to be 10.6 eV at the same level of theory, where HOMO is highest occupied molecular orbital and LUMO is lowest unoccupied molecular orbital. Since the gap between HOMO and LUMO is a good measure of molecular hardness, methane is a very hard molecule, which means that its ionization energy is very high and its electron affinity is very poor. As a consequence, it is very difficult to perform the selective activation of methane.

We have carried out mechanistic studies of the direct hydroxylation of methane by various enzymatic and catalytic systems at various levels of theory. In this chapter, the C–H bond activation and the subsequent hydroxylation of methane are considered on the basis of orbital interaction analyses and DFT calculations.

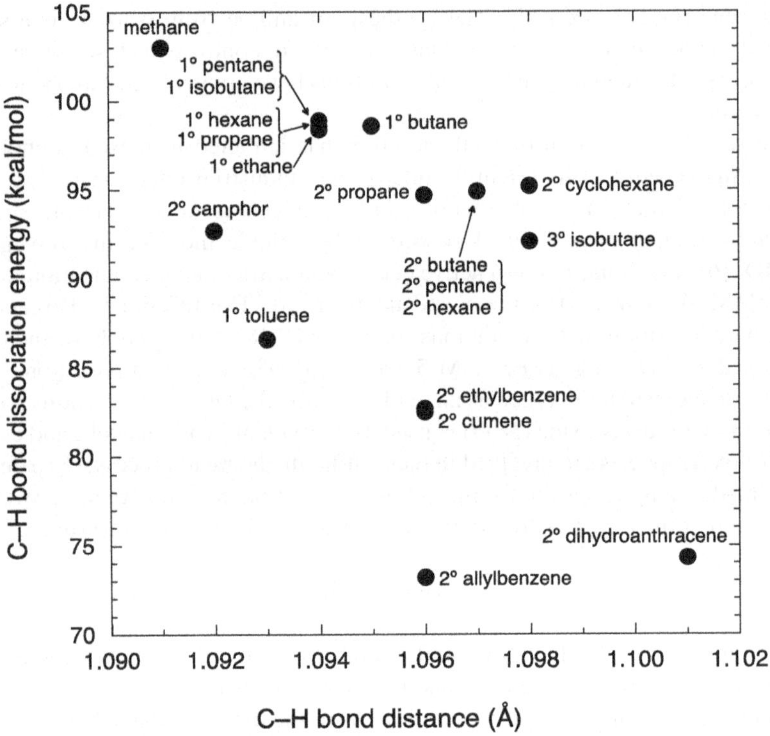

Fig. 1 C–H bond dissociation energies of alkanes as a function of C–H bond distance calculated at the B3LYP/6-311++G** level of theory

2 C–H Bond Activation of Methane

2.1 *Molecular Orbitals of Methane*

Let us first look at the molecular orbitals (MOs) of methane. The fragment molecular orbital (FMO) diagram shown in Fig. 3 tells us how the MOs of tetrahedral CH_4 are formed [10]. Since the 1s atomic orbital of the carbon atom deeply lies at −285 eV, we can reasonably neglect it for the formation of the MOs of methane. The H1s, C2s, and C2p orbitals closely lie at −13.6, −21.4, and −11.4 eV, respectively, as seen from tables of parameters for extended Hückel calculations. Therefore, the MOs of the cubic H_4 molecule shown at left of Fig. 3 and the atomic orbitals of the central carbon atom shown at right are allowed to mix to form the MOs of methane.

The a_1 and t_2 orbitals of the cubic H_4 fragment combine in-phase with the 2s atomic orbital and the $2p_x$, $2p_y$, and $2p_z$ atomic orbitals of the central carbon atom to form MOs $1a_1$ and $1t_2$, respectively. In a similar way, the a_1 and t_2 orbitals of the H_4 fragment combine out-of-phase with the 2s atomic orbital and the $2p_x$, $2p_y$, and $2p_z$ atomic orbitals of the carbon atom to form MOs $2a_1$* and $2t_2$*, respectively,

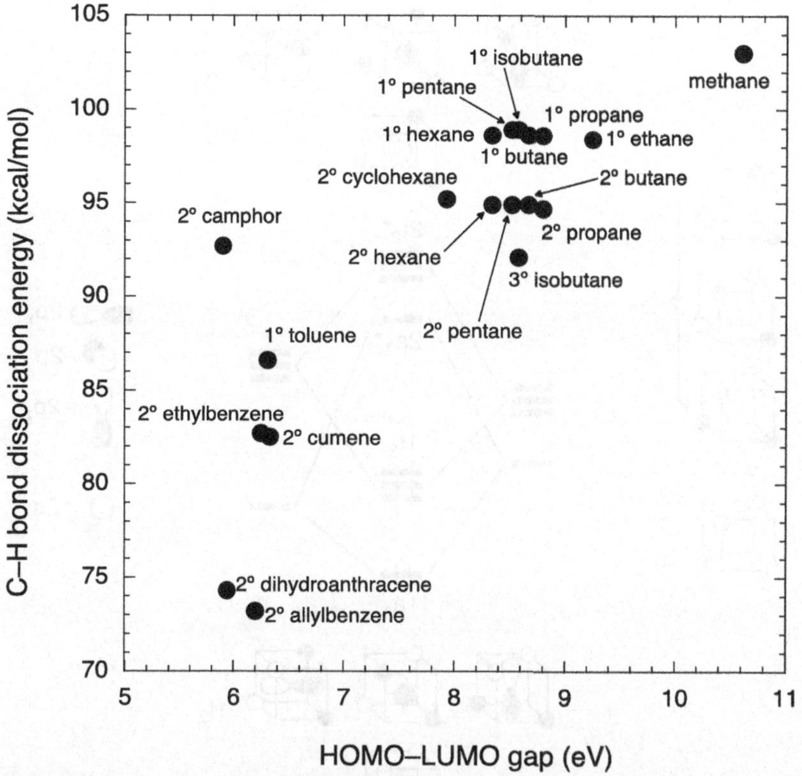

Fig. 2 C–H bond dissociation energies of alkanes as a function of HOMO–LUMO gap calculated at the B3LYP/6-311++G** level of theory

where symbol * shows antibonding MOs. It should be noted again that the threefold degenerate HOMOs of $1t_2$ come from the bonding combination of the t_2 MOs of the H_4 fragment and the three 2p orbitals of the carbon atom. The LUMO of $2a_1$* derives from the antibonding combination of the a_1 MO of the H_4 fragment and the 2s orbital of the carbon atom. The eight-electron CH_4 molecule is very stable in the tetrahedral structure, due to the full occupation of the low-lying $1a_1$ and threefold degenerate $1t_2$ MOs. The $1t_2$ HOMO and $2a_1$ LUMO have remarkable bonding and antibonding natures with respect to the C–H bonds, respectively, and thus, its HOMO–LUMO gap is large 10.6 eV at the B3LYP level of theory. Consequently, methane is a very hard molecule, and its C–H activation is difficult.

Table 2 lists these MO levels calculated by using extended Hückel, Hartree–Fock, and various DFT calculations. The results remind us again that the extended Hückel and Hartree–Fock methods tend to overestimate the HOMO–LUMO gaps of molecules, while the pure DFT methods (SVWN, BLYP, and PBE) tend to underestimate the gaps. Note that the $2a_1$* and $2t_2$* levels are reversed at the extended

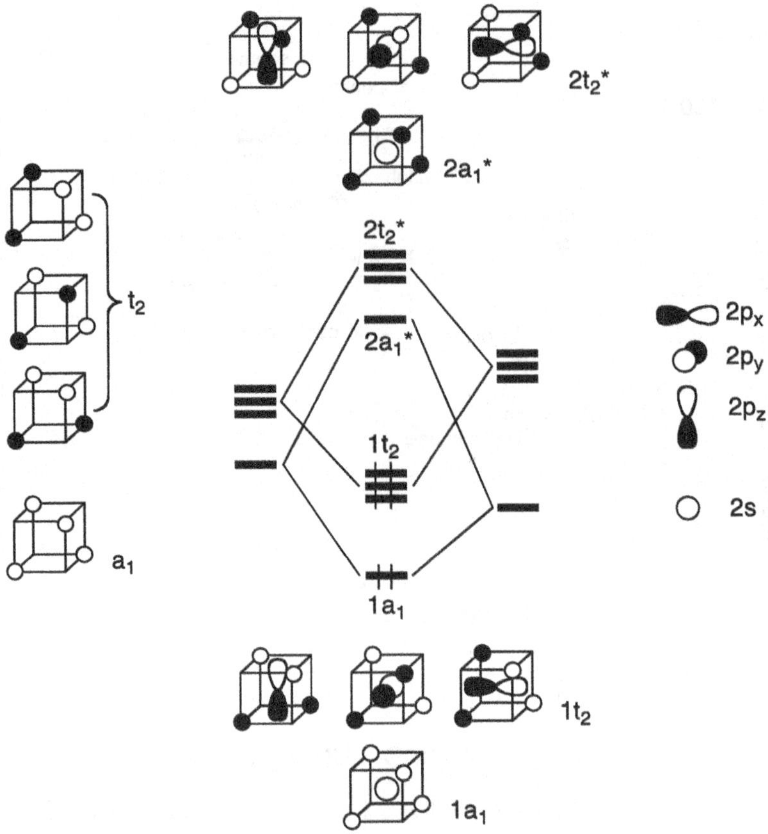

Fig. 3 Molecular orbitals of methane constructed from the cubic H_4 fragment (left) and the central carbon atom (right), where labels t are threefold degenerate orbitals

Hückel level of theory. We have considered and calculated how the C–H activation of methane and its hydroxylation take place by the specific enzyme systems and related metal-oxo species [11].

2.2 *Orbital Concept for Methane Activation Based on Second-Order Perturbation*

We consider that the C–H bond activation of methane can take place through the intermolecular interactions between methane and the active metal sites of the enzymes and catalysts. The strength of the interactions is determined by second-order perturbation, or orbital interaction, which gives a theoretical background of frontier orbital theory by Fukui and the Woodward–Hoffmann rules. The interaction energy $\Delta\epsilon_i$ is

Table 2 Computed molecular orbital energies of methane from extended Hückel calculations and Hartree–Fock and DFT calculations with the 6-311++G** basis set

Method		Molecular orbitals (eV)				HOMO–LUMO gap (eV)	Distance (Å)
		$1a_1$	$1t_2$	$2a_1{}^*$	$2t_2{}^*$		
Extended Hückel		−24.9	−15.6	37.0	4.9	20.4	1.091[a]
Hartree–Fock/6-311++G**		−25.7	−14.9	1.1	2.1	16.0	1.084
DFT	SVWN/6-311++G**	−17.4	−9.9	−0.6	0.8	9.3	1.097
	BLYP/6-311++G**	−16.9	−9.4	−0.4	0.9	8.9	1.097
	PBE/6-311++G**	−17.0	−9.4	−0.3	1.0	9.1	1.098
	B3LYP/6-311++G**	−19.0	−10.8	−0.2	1.1	10.6	1.091
	PBE0/6-311++G**	−21.7	−12.4	0.5	1.6	13.0	1.088
	M06/6-311++G**	−19.2	−11.0	−0.9	0.0	10.0	1.091
	TPSSh/6-311++G**	−18.4	−10.2	0.0	1.2	10.2	1.092

[a]Optimized by B3LYP

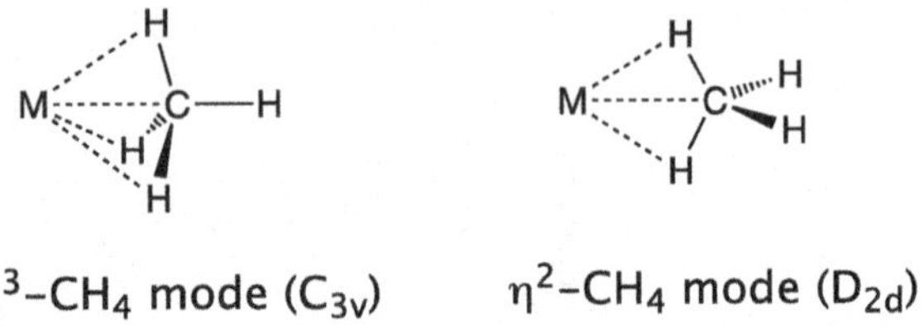

Fig. 4 Possible structures of five-coordinate methane, which is deformed into a C_{3v} or D_{2d} structure

given by using orbital $\psi_i^{(0)}$ of molecule A and orbital $\psi_j^{(0)}$ of molecule B, in which the corresponding energies are $\varepsilon_i^{(0)}$ and $\varepsilon_j^{(0)}$, respectively, on the basis of the following equation.

$$\Delta\varepsilon_i = \frac{\left|H_{ij}^{'}\right|^2}{\varepsilon_i^{(0)} - \varepsilon_j^{(0)}}$$

Here, H'_{ij} is the so-called resonance integral defined by the following equation.

$$H_{ij}^{'} = \left\langle \psi_i^{(0)} \middle| H' \middle| \psi_j^{(0)} \right\rangle$$

The magnitude of the interaction energy is governed by the orbital overlap (the magnitude of H_{ij} is roughly proportional to S_{ij}, the relevant overlap) and the energy difference between the two interacting orbitals. Our calculations indicate that a C_{3v}- or D_{2d}-distorted methane shown in Fig. 4 can be bound if the metal active site of the enzymes and catalysts involve a coordinatively unsaturated transition metal such as Fe and Cu. The methane complex is suggested to include a five-coordinate carbon species with an M–CH_4 bond. In the C_{3v} structure, the three H atoms that face the metal active center are opened, while in the D_{2d} structure, the two H atoms are opened, and at the same time, the two H atoms of the other side are also opened.

First, we have studied the importance of the C_{3v} deformation for methane (proposed by Shestakov and Shilov [12]) from the point of view of second-order perturbation theory [13]. We demonstrated from qualitative calculations that a C_{3v}-deformed methane can be theoretically activated on a supposed diiron active site of sMMO, if that site includes a five-coordinate iron active center. Our proposals include a complex with an interesting Fe–CH_4 bond in the initial stage of the catalytic cycles, just as in the left of Fig. 4. In the context of recent organometallic chemistry, a five-coordinate carbon species is not so unrealistic [14]. The previous orbital interaction analyses [13] have shown that interactions between the methane t_2 HOMO (C–H bonding) and the unoccupied d-block orbitals of ML_n model complexes (rather than the interactions between the LUMO (C–H antibonding) and the occupied d-block orbitals) play an essential role in activating the methane C–H bonds, as shown in Fig. 5. Our extended Hückel calculations suggested that a six-coordinate iron is not

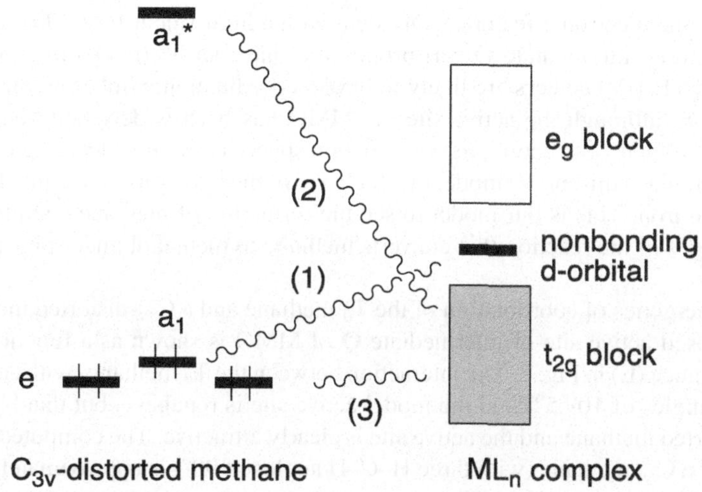

Fig. 5 Orbital interactions for a coordinatively unsaturated ML_n complex and C_{3v}-distorted methane

likely to possess direct reactivity with methane, even if it contains an active metal-oxo species. However, methane can be activated if a coordinatively unsaturated metal center, e.g., a five-coordinate metal is generated.

Let us look at the orbital interaction more in detail to understand essential features for methane activation. Figure 5 shows schematic orbital interaction drawing between a C_{3v}-distorted methane and a coordinatively unsaturated ML_n complex. The threefold degenerate t_2 HOMO of methane split into the a_1 and e orbitals, and as a consequence, the a_1 orbital is pushed up due to the geometrical change. When the ML_n complex has a five-coordinate metal, one of the e_g-block orbitals go down to the middle of the t_{2g} and e_g block orbitals. This is a nonbonding unfilled orbital. Since in general, two-electron-two-orbital interactions are always attractive, both interactions (1) and (2) should play an important role in the attractive interaction between the methane and the complex. However, interaction (1) is a major contributor compared to interaction (2) because the energy difference of interaction (1) is smaller than that of interaction (2). The four-electron-two-orbital interaction (3) causes a repulsive interaction, as known as steric repulsion in organic chemistry.

2.3 Methane Activation by sMMO Model

Let us next look at the interaction between methane and a simple sMMO model calculated by extended Hückel method. Que and Lipscomb and their co-workers [15] determined from EXAFS and Mössbauer analyses that intermediate **Q** has the structural and electronic features consistent with a high-valent $Fe_2(\mu\text{-}O)_2$ diamond core structure. Intermediate **Q** was proposed to have an $Fe_2(\mu\text{-}O)_2$ diamond core, and the two Fe(IV) centers are likely to have a coordination number no higher than 5. However, although the active site of sMMO has been widely believed to have the bis(μ-oxo)Fe^{IV}_2 structure, its coordination sphere is still not clear at present. We considered that a diiron(IV) model in which one of the two iron centers involve a five coordinate iron. This is our model to see the formation of methane complex at the initial stages of the reaction that converts methane to methanol under physiological conditions.

The energetics of coordination of the T_d methane and a C_{3v}-distorted methane to the supposed active site of intermediate **Q** of MMO is shown as a function of the Fe–C distance (L) in Fig. 6. The interaction between the T_d methane (with equivalent H–C–H angles of 109.5°) and the model active site is repulsive, but that between a C_{3v}-distorted methane and the active site is clearly attractive. The computed binding energy of a C_{3v} methane (with three H–C–H angles of 90°) is approximately 0.1 eV in this model (at $L = 2.3$ Å). This is a clear indication of the attractive interaction. Therefore, a C_{3v}-deformed methane can be theoretically bound at a supposed diiron

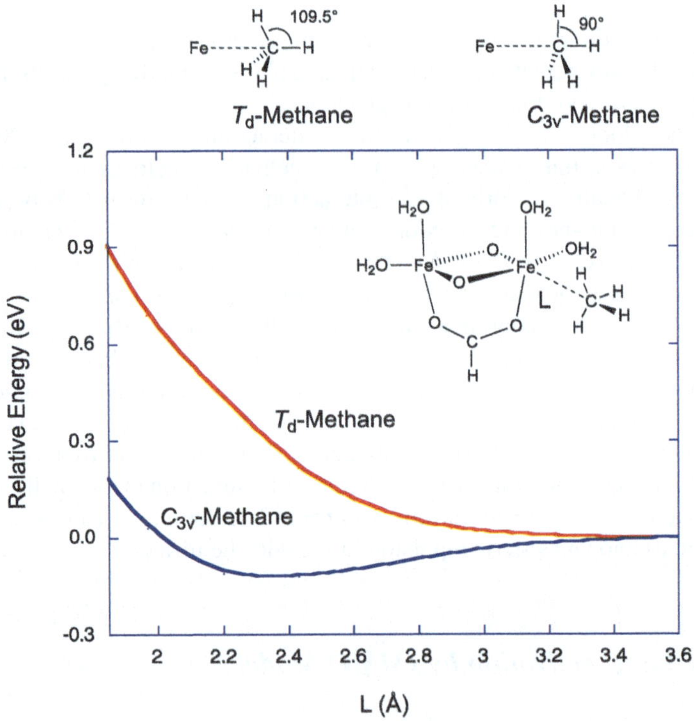

Fig. 6 Relative energies of T_d and C_{3v} methane and an sMMO model with five-coordinate iron calculated with the extended Hückel method. Reproduced with permission from Ref. [13]. Copyright 1997 American Chemical Society

active site of MMO, if that site has a five-coordinate iron active center in it. Our proposals include a methane complex with an interesting Fe–CH_4 bond (Fig. 4, left) in the initial stages of the catalytic reaction.

The coordination of a C_{3v}-distorted methane that we have discussed above is one possibility, denoted as an η^3-binding mode. There is another possible way of coordination for methane, which we call an η^2-binding mode, as shown in the right of Fig. 4. The two binding modes were found in our more reliable calculations on the $FeO^+(CH_4)$ complex, actually both minima on a potential energy surface from DFT calculations [16, 17].

Let us next look at the interaction between a D_{2d}-distorted methane and a model of intermediate **Q** of sMMO. The total energy diagrams for the coordination of the T_d methane and D_{2d}-distorted methane to the supposed active site are shown in Fig. 7. The interaction between a D_{2d}-distorted methane and the active site is attractive also in this binding mode, but that between the T_d methane and the active site is always repulsive. The computed binding energy of a D_{2d} methane (with two H–C–H angles opened up to 150°) is about 0.15 eV in this model (at $L = 2.4$ Å), consistent with the calculational result based on a different model with an η^3-binding mode (C_{3v}

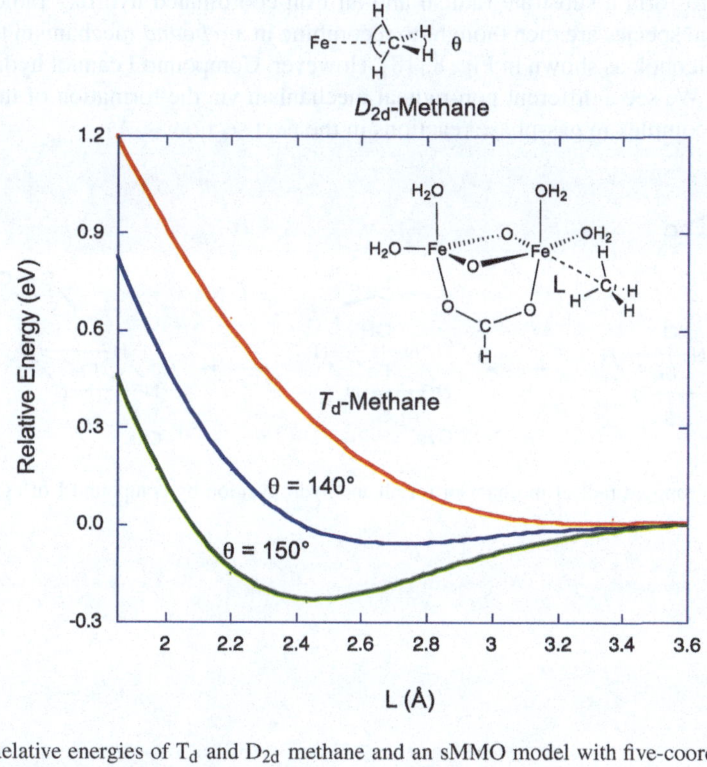

Fig. 7 Relative energies of T_d and D_{2d} methane and an sMMO model with five-coordinate iron calculated with the extended Hückel method. Reproduced with permission from Ref. [13]. Copyright 1997 American Chemical Society

methane) in Fig. 6. Methane can be thus bound at the five-coordinate iron center of a model of intermediate **Q** both in the η^2-binding mode as well as in the η^3-binding mode.

Thus, we think that a five-coordinate iron would activate methane in a similar manner. The C–H bonds of methane are activated in these complexes because of significant electron transfer from the methane t_2 HOMO (C–H bonding) to the unfilled nonbonding d orbital of the model iron complex through interaction (1), as shown in Fig. 5. In fact, the computed total charge of the methane is +0.18 in the complex with an η^3-binding mode in Fig. 6 (at $L = 2.3$ Å) and +0.28 in the complex with an η^2-binding mode in Fig. 7 (at $L = 2.4$ Å).

Our proposal argues differ from the so-called oxygen rebound mechanism widely accepted in bioinorganic chemistry. For instance, in a proposed catalytic mechanism for cytochrome P-450 (which is able to hydroxylate a variety of secondary (2°) and tertiary (3°) C–H bonds of alkanes) [18], the O–O bond is cleaved at the heme iron to produce a diatomic unit which may be written as Fe(III)–O^0, Fe(IV)–O^-, or Fe(V)–O^{2-}. The π system of the porphyrin ring may also be oxidized to a cation radical; as a result, the so-called Compound I, with an Fe(III)–O^- or Fe(IV) = O^{2-} core structure, is formed. This species is postulated to directly abstract a hydrogen atom from hydrocarbons, to form a substrate radical and an iron-coordinated hydroxy radical. The two radical species are then thought to recombine in a *rebound* mechanism to afford product alcohol, as shown in Fig. 8 [18]. However, Compound I cannot hydroxylate methane. We see a different non-radical mechanism via the formation of the initial methane complex in gas-phase reactions in the next section.

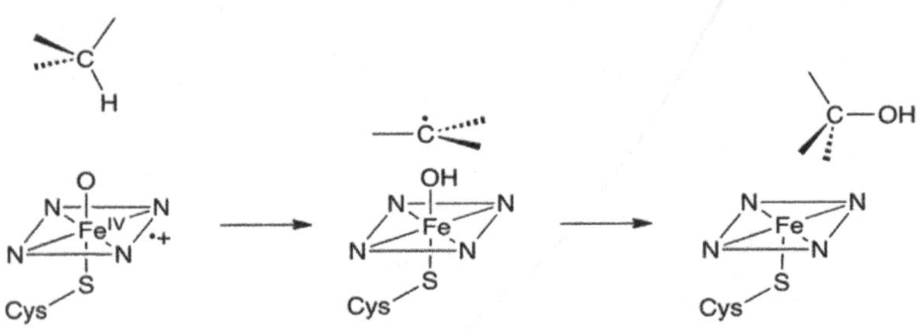

Fig. 8 A proposed radical mechanism for alkane hydroxylation by compound I of cytochrome P450

3 Reaction Mechanism for the Direct Hydroxylation of Methane

3.1 Methane Hydroxylation by the FeO$^+$ Species

The transition-metal-oxide ions (MO$^+$s) are excellent oxidants. Gas-phase reactions between MO$^+$s and hydrocarbons are of particular interest since they can be viewed as good model reactions for various oxidation reactions by catalytic and enzymatic systems. Methane hydroxylation by MO$^+$s in the gas phase under ion-cyclotron-resonance conditions has been investigated by Schwarz and coworkers [3, 8] and Armentrout and coworkers [19]. The catalytic activity of the bare MO$^+$ complexes toward methane is a key to the mechanistic aspects in the direct methane hydroxylation [8, 19–27].

Schröder, Schwarz, and co-workers have systematically investigated the gas-phase reactions of the first-row MO$^+$ complexes and methane using the Fourier-transformed mass spectroscopic analysis under ion cyclotron resonance conditions [3, 8, 24–26]. They demonstrated that the late MO$^+$ complexes are able to activate methane while the early ones are not. The reaction efficiency and the methanol branching ratio are significantly dependent on metals. For example, FeO$^+$ efficiently reacts with methane, forming methanol in 41% yield [8]. Although MnO$^+$ reacts with methane very efficiently, the branching ratio to methanol is less than 1% [25]. CoO$^+$ exhibits low reactivity toward methane, but the branching ratio to methanol is 100% [19]. Both the reactivity and the methanol branching ratio are high in NiO$^+$ [3]. In contrast, the early MO$^+$ complexes (ScO$^+$, TiO$^+$, and VO$^+$) exhibit no reactivity toward alkanes and alkenes, due to their strong metal–oxo bonds. Interestingly, Sc$^+$ reacts with methanol to yield ScO$^+$ and methane in the gas phase [28], which is precisely the reverse reaction of methane hydroxylation.

On the basis of detailed DFT calculations, we showed that there are two possible reaction pathways for methane hydroxylation, as indicated in Fig. 9 [16, 29]. In the initial stages of the reaction pathway, an interesting methane complex is formed

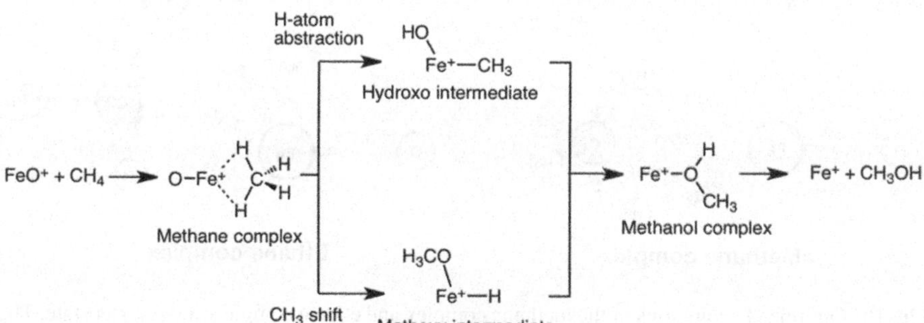

Fig. 9 Two possible reaction pathways for methane hydroxylation by FeO$^+$

in prior to C–H bond activation. Direct or indirect observations have been made to support the intermediacy of alkane complexes in C–H activation [30], but their direct observation requires ultrafast spectroscopic techniques at very low temperature. The initially formed methane complex $FeO(CH_4)^+$, which takes an η^2-CH_4 coordination, is transformed into hydroxo intermediate ($HO–Fe–CH_3^+$) and methoxy intermediate ($H–Fe–OCH_3^+$) by an H-atom abstraction and methyl shift, respectively. The reaction pathway via hydroxo intermediate is energetically more favorable than the other one via methoxy intermediate [16]. This result is fully consistent with the experimental prediction [8] that hydroxo intermediate should play a central role as an intermediate in the gas-phase reaction between FeO^+ and methane. Hydroxo intermediate is then transformed into the methanol complex as a result of the recombination of the OH and CH_3 ligands at the metal center.

Detailed discussion about this reaction pathway is developed in Chapter "Theoretical Study of the Direct Conversion of Methane by First-Row Transition-Metal Oxide Cations in the Gas Phase". We show that the methane hydroxylation by the FeO^+ species should take place in the sextet and quartet states along the reaction pathway via the hydroxo intermediate [29]. Thus, this reaction can be viewed as a spin-crossover reaction. The structure of methane in the initial complex is significantly deformed from the T_d-type structure; the H–C–H angle of the coordination side is deformed from 109.5° to 120°. A D_{2d}-type distortion of methane is actually observed from a geometrical optimization of the methane complex $FeO(CH_4)^+$, as shown in Fig. 10. The interactions between the HOMO of the coordinated methane (C–H bonding) and the unfilled orbitals of FeO^+ and between the LUMO of the methane (C–H antibonding) and the filled orbitals of FeO^+ play an essential role in the formation of this complex. One of the hydrogen atoms in the coordinated methane shifts to the oxygen atom via a four-centered transition state (TS1) to generate the hydroxo intermediate, as shown in Fig. 11. In the second half of the reaction, a recombination occurs to form a C–O bond via a three-centered transition state (TS2),

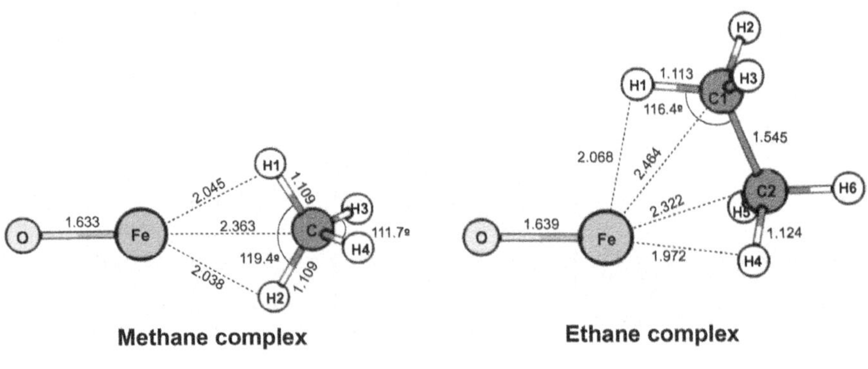

Fig. 10 Optimized geometries of the methane complex and ethane complex in the sextet state. The units are in Å

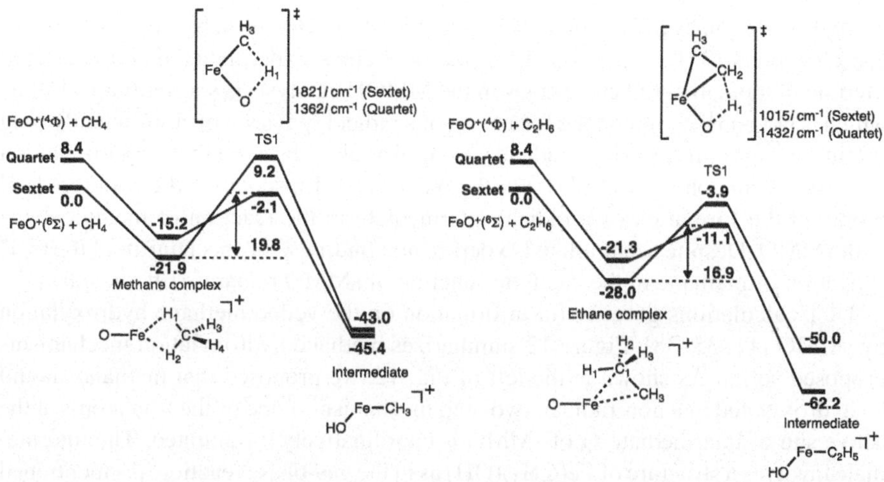

Fig. 11 Reaction energy diagrams for the C–H abstraction reactions of methane and ethane by FeO^+ in units of kcal/mol. The numbers about the transition states are calculated imaginary frequencies

leading to the methanol complex, $Fe(CH_3OH)^+$. This process occurs at a coordinatively unsaturated metal active center, so no radical species is involved in the present mechanism.

Let us look at the energetics for the C–H abstraction reactions of methane and ethane by FeO^+ shown in Fig. 11. The reactions proceed from the formation of methane and ethane complexes. The ground state of FeO^+ is spin sextet, whereas that of TS1 is spin quartet. Therefore, spin inversion takes place in the course of the reactions. As a result of the spin inversion, TS1 is significantly decreased in energy. The activation energies of TS1 measured from the initial complexes are computed to be 19.8 and 16.9 kcal/mol for methane and ethylene, respectively. The energy difference of 3 kcal/mol should have a significant difference in the reaction rate for the C–H cleavage of methane and ethane. On the basis of transition state theory, we can predict that the reaction rate of methane is approximately 100 times slower than that of ethane. This DFT result is fully consistent with the C–H bond dissociation energies of methane and ethane, as shown in Figs. 1 and 2.

3.2 *Methane Hydroxylation Mechanisms by sMMO*

As mentioned above, Que, Lipscomb, and coworkers [15] showed from a combined Mössbauer–EXAFS investigation that the active site of intermediate **Q** should involve a bis(μ-oxo)diiron^{IV} core, in which the two iron atoms are antiferromagnetically coupled [31]. The EXAFS study suggested that the coordination number of the iron atoms should be no greater than 5. One mechanism for the hydroxylation by intermediate **Q** is a radical rebound mechanism, which is widely believed to occur in the hydroxylation by cytochrome P450; see Fig. 8 [18]. However, Newcomb, Lippard, and coworkers [32] demonstrated from radical clock experiments that a measured lifetime of a putative radical species in the MMOH catalysis is shorter than ~150 fs, which is not consistent with the formation of a radical species with a sufficiently long lifetime. A short lifetime for a radical species was also observed in the hydroxylation of chiral ethane on sMMO [33]. On the other hand, Lipscomb and coworkers [34] proposed the formation of a radical intermediate in the reaction of methylcubane with sMMO. Despite accumulated experimental findings, the mechanism of the C–H activation of methane in the catalytic function of sMMO remains still unclear.

DFT calculations gave useful information on the veiled methane hydroxylation by sMMO [11, 35–38]. Figure 12 summarizes methane hydroxylation mechanisms proposed so far. As shown at the left of Fig. 12, we proposed that methane should be hydroxylated in a non-radical, two-step mechanism if one of the iron atoms at the active site of intermediate **Q** of sMMO is coordinatively unsaturated. The intermediate involves a structure of $Fe(CH_3)(OH)$ as in the gas-phase reaction. As mentioned above in Fig. 11, intermediate $HO–Fe–CH_3$ is involved in the methane hydroxylation reaction by FeO^+ in the gas phase.

There are other mechanistic proposals for methane hydroxylation by **Q**. As shown at the center (left) of Fig. 12, Siegbahn and Crabtree proposed using a five-coordinate iron model with high-spin nonet and undecet states that the methyl radical should recombine with an iron center via a weak $Fe–CH_3$ bond after the H-atom abstraction [36]. This mechanism is somewhat similar to our proposal in that methyl radical

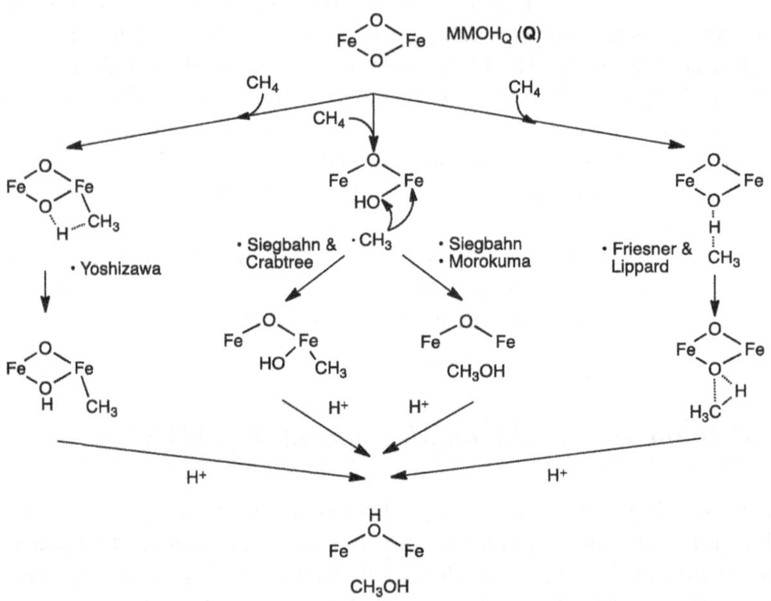

Fig. 12 Mechanisms for methane hydroxylation by intermediate **Q** of sMMO by DFT calculations. Reproduced with permission from Ref. [11]. Copyright 2006 American Chemical Society

is trapped during the reaction. On the other hand, as shown at the center (right), Siegbahn [36, 36] and Morokuma and coworkers [37] proposed that a C–H bond of methane is cleaved in a homolytic manner by six-coordinate diiron model complexes in the high-spin states. After the dissociation of a C–H bond of methane, the resultant methyl radical is shifted to the formed OH group. These authors proposed that **Q**-mediated hydroxylation should proceed along the radical rebound mechanism of cytochrome P450.

In contrast to the radical mechanism, Friesner, Lippard, and coworkers [38] proposed a non-synchronous concerted mechanism, as indicated at the right of Fig. 12. Since the C–H activation and the rebound process depend on the coordination sphere and spin state of diiron models adopted, it is important to use an appropriate diiron model of **Q**. Detailed discussion about the mechanism of sMMO is developed in Chapter "Enzymatic Methane Hydroxylation: sMMO and pMMO".

3.3 Methane Hydroxylation Mechanisms by pMMO

The structure of pMMO determined to 2.8 Å resolution shows a trimeric arrangement and overall folds of three subunits [39]. There are three metal centers per protomer in the crystal structure. Two of these, which were modeled as mononuclear and dinuclear copper species, are located within the soluble regions of the pmoB subunit, the two copper sites being 21 Å apart from each other. In a previous DFT study [35], we proposed a mechanism for the C–H cleavage and the recombination between CH_3 and OH ligands using a simple mixed-valent dinuclear $Cu^{II}Cu^{III}$ cluster with ammonia and hydroxo ligands. This $(\mu\text{-O})_2Cu^{II}Cu^{III}$ species has good power for methane hydroxylation compared with the μ-η^2:η^2-peroxo$Cu^{II}Cu^{II}$ and $(\mu\text{-O})_2Cu^{III}Cu^{III}$ species, which show no reactivity for the activation of methane.

We reported the mechanisms of methane hydroxylation at the mononuclear and dinuclear copper sites on the basis of the crystal structure of pMMO and considered how reactive copper species are formed in the protein environment [40, 41]. By looking at the coordination environments, we set up three kinds of model complexes for DFT calculations to search the reaction pathway for the conversion of methane to methanol. One is a mononuclear Cu^{III}–O (or Cu^{II}–O$^{\cdot}$) model with two imidazole and one acetate, and others are mixed-valent $(\mu\text{-O})_2Cu^{II}Cu^{III}$ and $(\mu\text{-O})(\mu\text{-OH})Cu^{II}Cu^{III}$ models with three imidazole and one acetate.

Figure 13 shows a mechanism for methane hydroxylation by the monocopper-oxo species. The copper-oxo species optimized by using a small model in the gas phase is in good agreement with the one optimized in the protein environment with respect to the coordination bonds around the central copper atom. In the initial stages of the reaction, methane is weakly bound to the monocopper active center, and after that, one of the C–H bonds of methane is cleaved by the oxo species via the first transition state (TS1).

The resultant radical species leads to a non-radical intermediate, which is extremely stable in energy. Since the formal charge of the copper ion is changed from

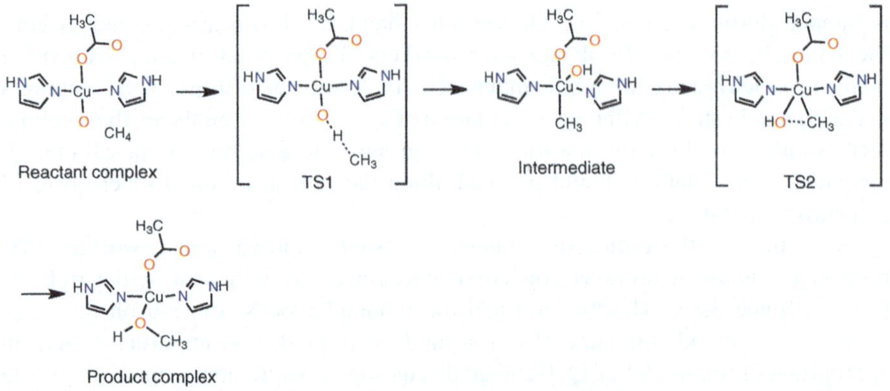

Fig. 13 Possible mechanism for the conversion of methane to methanol by a Cu^{III}–O (or Cu^{II}–O$^{\cdot}$) species of pMMO at the B3LYP level of theory

+3 to +1 after the H-atom abstraction, this result about a d^{10} system is reasonable. The second transition state (TS2) has a typical triangle structure for the recombination between the OH and CH_3 ligands. This non-radical mechanism is identical to the ligand-coupling mechanism of Barton's *Gif* chemistry [42]. We expect that the spin crossover from triplet to singlet should take place in the course of this reaction in the vicinity of the H-atom abstraction step. The general features of this reaction are similar to those of methane hydroxylation by the bare CuO^+ complex [16, 29, 43] and of dopamine hydroxylation by a possible copper-oxo species of dopamine β-monooxygenase [44, 45].

As discussed previously [35, 40, 41], we concluded that the peroxo species and the $(\mu\text{-O})_2Cu^{III}Cu^{III}$ species have no direct ability to activate methane, while the $(\mu\text{-O})_2Cu^{II}Cu^{III}$ species can activate the inert C–H bond methane. Here, let us take a look at the mechanism of the $(\mu\text{-O})(\mu\text{-OH})Cu^{II}Cu^{III}$ species with respect to the hydroxylation of methane. Figure 14 shows the reaction pathway starting from the dissociation limit to the final complex that involves the tyrosine residue, the methanol product, and the $(\mu\text{-O})Cu^{II}Cu^{II}$ species in the triplet state and singlet state [46].

The hydroxylation of methane by $(\mu\text{-O})(\mu\text{-OH})Cu^{II}Cu^{III}$ starts with the formation of a methane complex. The first transition state TS1 leads to the C–H bond dissociation of methane. The H-atom abstraction forms a methyl intermediate; no radical species is formed in this mechanism. This mechanistic proposal is consistent with the experimental observation [47] that chiral ethane hydroxylation by pMMO from *Methylococcus capsulatus* (Bath) exhibits negligible racemization.

The mechanism connects the intermediate and the methanol complex via the second transition state TS2 in the C–O bond formation step. In the final step, the migrated H-atom returns to the phenoxyl radical of tyrosine from the bridging hydroxo ligand, resulting in the formation of the final complex corresponding to a complex of the tyrosine residue, the methanol product, and the $(\mu\text{-O})Cu^{II}Cu^{II}$

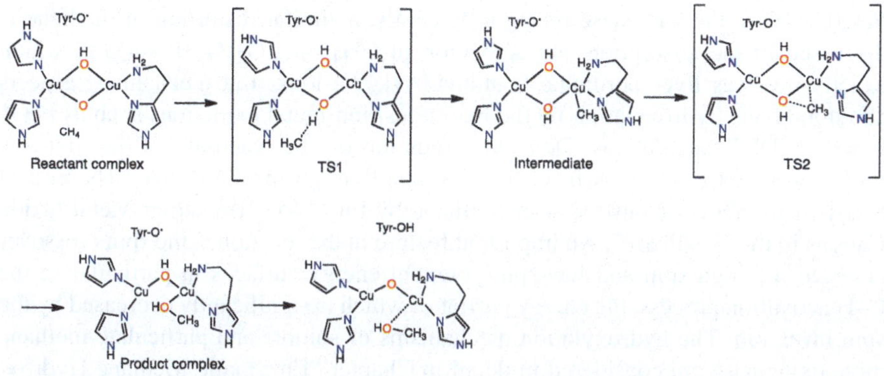

Fig. 14 Possible mechanism for the conversion of methane to methanol by the (μ-O)(μ-OH)$Cu^{II}Cu^{III}$ species of pMMO, in which the tyrosine residue located in the second-coordination sphere plays a role in the formation of the active species

species. Since the calculated activation energy of TS2 is lower than TS1, the rate-determining step is the C–H bond dissociation in the methane hydroxylation. These calculated activation barriers are consistent with experimental KIE (k_H/k_D) values of 5.2–5.5 in ethane hydroxylation by pMMO, which suggests that the rate-determining step is the C–H bond activation step [47].

We have considered the reactivity of the mono- and dinuclear Cu sites of pMMO using DFT calculations, but the resolution of X-ray analyses [39], from which we constructed our calculation models, was low and the geometry of the dinuclear site a little unusual. Recently, Rosenzweig, Ryde, and their coworkers have done quantum refinement (crystallographic refinement enhanced with quantum chemical calculations) to improve the structure of the active site [48]. They reported that the best results were obtained with mononuclear Cu sites, occasionally with an extra water molecule. We therefore need to pay more attention to the monocopper site for methane hydroxylation.

4 Conclusions

In this chapter, we have reviewed our theoretical studies on methane C–H bond activation and hydroxylation by soluble and particulate methane monooxygenase (iron and copper enzyme species) and related bare transition-metal-oxo ions such as FeO^+. We propose that in the initial stages of the reaction, the tetrahedral T_d structure of methane should be deformed into a C_{3v} or D_{2d} structure at coordinatively unsaturated metal-oxo species. Methane can be bound from extended Hückel calculations at a five-coordinate iron center of a model of intermediate **Q** both in the η^3-binding mode (C_{3v}) and in the η^2-binding mode (D_{2d}). Actually, a D_{2d}-type distortion of methane is obtained from DFT calculations in the methane complex

$FeO(CH_4)^+$ in the gas-phase reaction. Recently, a similar distortion of methane is discussed in the adsorption and activation of CH_4 on IrO_2 (110) surface, which exactly involves five-coordinate Ir atoms [49]. We looked at mechanistic aspects about methane hydroxylation by the bare transition-metal oxide ions such as FeO^+ based on DFT calculations. Detailed discussion on the reactivity of the first-row transition-metal oxide ions from ScO^+ to CuO^+ is given in Chapter "Theoretical Study of the Direct Conversion of Methane by First-Row Transition-Metal Oxide Cations in the Gas Phase". An important feature in the reaction is the spin crossover between the high-spin and low-spin potential energy surfaces in particular in the C–H activation process, the energy barrier of which is significantly decreased by the spin inversion. The hydroxylation mechanisms of soluble and particulate methane monooxygenase are considered in detail in Chapter "Enzymatic Methane Hydroxylation: sMMO and pMMO". These mechanistic insights are reasonably extended to the things in metal-exchanged zeolites in Chapters "Mechanistic Understanding of Methane Hydroxylation by Cu-Exchanged Zeolites" and "Oxidative Activation of Metal-Exchanged Zeolite Catalysts for Methane Hydroxylation" and IrO_2 and β-PtO_2 (110) surfaces in Chapter "Dynamics and Energetics of Methane on the Surfaces of Transition Metal Oxides".

Acknowledgements K.Y. gives special thanks to JST-CREST "Innovative Catalysts" Grant No. JPMJCR15P5 for its support of our theoretical project about natural-gas utilization. K.Y. acknowledges KAKENHI from Japan Society for the Promotion of Science (JSPS) and the Ministry of Education, Culture, Sports, Science and Technology of Japan (MEXT) and also thanks the MEXT Projects of "Integrated Research Consortium on Chemical Sciences," "Cooperative Research Program of Network Joint Research Center for Materials and Devices," and "Elements Strategy Initiative to Form Core Research Center". The computations were performed at the Research Institute for Information Technology (Kyushu University).

References

1. (a) Shilov AE (1984) Activation of saturated hydrocarbons by transitions metal complexes. Reidel, Dordrecht (b) Shilov AE, Shul'pin GB (1997) Chem Rev 97:2879–2932
2. (a) Crabtree RH (1990) The organometallic chemistry of the transition metals. Wiley, New York (b) Crabtree RH (1995) Chem Rev 95:987–1007
3. (a) Schröder D, Schwarz H (1995) Angew Chem Int Ed 34:1973–1995 (b) Schwarz H, Schröder D (2000) Pure Appl Chem 72:2319–2332
4. Arndtsen BA, Bergman RG, Mobley TA, Peterson TH (1995) Acc Chem Res 28:154–162
5. Stahl SS, Labinger JA, Bercaw JE. (1998) Angew Chem Int Ed 37:2180–2192
6. (a) Feig AL, Lippard SJ (1994) Chem Rev 94:759–805 (b) Wallar BJ, Lipscomb JD (1996) Chem Rev 96: 2625–2657 (c) Que L, Dong Y (1996) Acc Chem Res 29:190–196 (d) Merkx M, Kopp DA, Sazinsky MH, Blazyk JL, Müller J, Lippard SJ (2001) Angew Chem Int Ed 40:2782–2807
7. (a) Lieberman RL, Shrestha DB, Doan PE, Hoffman BM, Stemmler TL, Rosenzeig AC (2003) Proc Natl Acad Sci USA 100:3820–3825 (b) Chan SI, Chen KH-C, Yu SS-F, Chen C-Li, Kuo SS-J (2004) Biochemistry 43:4421–4430
8. (a) Schröder D, Schwarz H (1990) Angew Chem Int Ed Engl 29:1433–1434 (b) Schröder D, Fiedler A, Hrusák J, Schwarz H (1992) J Am Chem Soc 114:1215–1222

9. (a) Anderson JR, Tsai P (1987) J Chem Soc Chem Commun 1435–1436 (b) Sobolev VI, Dubkov KA, Panna OV, Panov GI (1995) Catal Today 24:251–252 (c) Dubkov KA, Sobolev VI, Talsi EP, Rodkin MA, Watkins NH, Shteinman AA, Panov GI (1997) J Mol Catal A Chemical 123:155–161
10. Yoshizawa K (2001) J Organomet Chem 635:100–109
11. Reviews (a) Yoshizawa K (2006) Acc Chem Res 39:375–382 (b) Yoshizawa K (2013) Bull Chem Soc Jpn 86:1083–1116
12. Shestakov AF, Shilov AE (1996) J Mol Cat A 105:1–7
13. (a) Yoshizawa K, Yamabe T, Hoffmann R (1997) New J Chem 21:151–161 (b) Yoshizawa K, Ohta T, Yamabe T, Hoffmann R (1997) J Am Chem Soc 119:12311–12321
14. Spence REvonH, Parks DJ, Piers WE, MacDonald MA, Zaworotko MJ, Rettig SJ (1995) Angew Chem Int Ed Engl 34:1230–1233
15. Shu L, Nesheim JC, Kauffmann K, Münck E, Lipscomb JD, Que L (1997) Science 275:515–518
16. Yoshizawa K, Shiota Y, Yamabe T (1997) Chem Eur J 3:1160–1169
17. Yoshizawa K, Shiota Y, Yamabe T (1998) Organometallics 17:2825–2831
18. (a) Groves JT (1985) J Chem Educ 62:928–931 (b) Ortiz de M, Paul R (1995) Cytochrome P450: Structure, Mechanism, and Biochemistry, 2nd ed, Plenum, New York
19. Chen Y-M, Clemmer DE, Armentrout PB (1994) J Am Chem Soc 116:7815–7826
20. Jackson TC, Jacobson DB, Freiser BS (1984) J Am Chem Soc 106:1252–1257
21. Kang H, Beauchamp JL (1986) J Am Chem Soc 108:5663–5668
22. Kang H, Beauchamp JL (1986) J Am Chem Soc 108:7502–7509
23. Jackson TC, Carlin TJ, Freiser BS (1986) J Am Chem Soc 108:1120–1126
24. Ryan MF, Fiedler A, Schröder D, Schwarz H (1994) Organometallics 13:4072–4081
25. Ryan MF, Fiedler A, Schröder D, Schwarz H (1995) J Am Chem Soc 117:2033–2040
26. Schwarz J, Wesendrup R, Schröder D, Schwarz H (1996) Chem Ber 129:1463–1474
27. Clemmer DE, Aristov N, Armentrout PB (1993) J Phys Chem 97:544–552
28. Azzaro M, Breton S, Decouzon M, Geribaldi S (1993) Int J Mass Spectrom Ion Processes 128:1–20
29. Yoshizawa K, Shiota Y, Yamabe T (1998) J Am Chem Soc 120:564–572
30. Hall C, Perutz RN (1996) Chem Rev 96:3125–3146
31. Lee S-K, Fox BG, Froland WA, Lipscomb JD, Münck EA (1993) J Am Chem Soc 115:6450–6451
32. (a) Liu KE, Johnson CC, Newcomb M, Lippard SJ (1993) J Am Chem Soc 115:939–947 (b) Choi SY, Eaton PE, Hollenberg PF, Liu KE, Lippard SJ, Newcomb M, Putt DA, Upadhyaya SP, Xiong Y (1996) J Am Chem Soc 118:6547–6555
33. Valentine AM, Wilkinson B, Liu KE, Komar-Panicucci S, Priestley ND, Williams PG, Morimoto H, Floss HG, Lippard SJ (1997) J Am Chem Soc 119:1818–1827
34. (a) Jin Y, Lipscomb JD (1999) Biochemistry 38:6178–6186 (b) Brazeau BJ, Austin RN, Tarr C, Groves JT, Lipscomb JD (2001) J Am Chem Soc 123:11831–11837
35. (a) Yoshizawa K (1998) J Biol Inorg Chem 3:318–324 (b) Yoshizawa K, Ohta T, Yamabe T (1998) Bull Chem Soc Jpn 71: 1899–1909 (c) Yoshizawa K, Suzuki A, Shiota Y, Yamabe T (2000) Bull Chem Soc Jpn 73:815–827 (d) Yoshizawa K (2000) J Inorg Biochem 78:23–34 (e) Yoshizawa K, Yumura T (2003) Chem Eur J 9:2347–2358
36. (a) Siegbahn PEM, Crabtree RH (1997) J Am Chem Soc 119:3103–3113 (b) Siegbahn PEM, Crabtree RH, Nordlund P (1998) J Biol Inorg Chem 3:314–317 (c) Siegbahn PEM (1999) Inorg Chem 38:2880–2889 (d) Siegbahn PEM, Blomberg MRA (2000) Chem Rev 100:421–438 (e) Siegbahn PEM (2001) J Biol Inorg Chem 6:27–45
37. (a) Basch H, Mogi K, Musaev DG, Morokuma K (1999) J Am Chem Soc 121:7249–7256 (b) Basch H, Musaev DG, Mogi K, Morokuma K (2001) J Phys Chem A 105:3615–3622 (c) Basch H, Musaev DG, Morokuma K (2001) J Phys Chem B 105:8452–8460 (d) Torrent M, Musaev DG, Basch H, Morokuma K (2002) J Comput Chem 23:59–76
38. (a) Dunietz BD, Beachy MD, Cao Y, Whittington DA, Lippard SJ, Friesner RA (2000) J Am Chem Soc 122:2828–2839 (b) Gherman BF, Dunietz BD, Whittington DA, Lippard SJ, Friesner RA (2001) J Am Chem Soc 123:3836–3837 (c) Friesner RA, Dunietz BD (2001) Acc Chem Res 34:351–358 (d) Guallar V, Gherman BF, Miller WH, Lippard SJ, Friesner RA (2002) J Am Chem Soc 124:3377–3384

39. (a) Lieberman RL, Rosenzweig AC (2005) Nature 434:177–182 (b) Smith SM, Rawat S, Telser J, Hoffman BM, Stemmler TL, Rosenzweig AC (2011) Biochemistry 269:26344–26348 (c) Sirajuddin S, Barupala D, Helling S, Marcus M, Stemmler TL, Rosenzweig AC (2014) J Biol Chem 289:21782–21794
40. Yoshizawa K, Shiota Y (2006) J Am Chem Soc 128:9873–9881
41. Shiota Y, Yoshizawa K (2009) Inorg Chem 48:838–845
42. Barton DHR, Doller D (1992) Acc Chem Res 25:504–512
43. Shiota Y, Yoshizawa K (2000) J Am Chem Soc 122:12317–12326
44. Kamachi T, Kihara N, Shiota Y, Yoshizawa K (2005) Inorg Chem 44:4226–4236
45. Yoshizawa K, Kihara N, Kamachi T, Shiota Y (2006) Inorg Chem 45:3034–3041
46. (a) Shiota Y, Juhász G, Yoshizawa K (2013) Inorg Chem 52: 7907–7917 (b) Miyanishi M, Abe T, Hori Y, Shiota Y, Yoshizawa K (2019) Inorg Chem 58:12280–12288
47. Wilkinson B, Zhu M, Priestley ND, Nguyen H-HT, Morimoto H, Williams PG, Chan SI, Floss HG (1996) J Am Chem Soc 118:921–922
48. Cao L, Caldararu O, Rosenzweig AC, Ryde U (2018) Angew Chem Int Ed 57;162–166
49. Tsuji Y, Yoshizawa K (2018) J Phys Chem C 122: 15359–15381

Theoretical Study of the Direct Conversion of Methane by First-Row Transition-Metal Oxide Cations in the Gas Phase

Yoshihito Shiota and Kazunari Yoshizawa

Abstract The direct conversion of methane to methanol is a thermodynamically favorable process compared with the available commercial process that uses synthesis gas. This chapter reviews quantum chemical approaches, especially by means of density functional theory (DFT) calculations, to elucidate the reaction pathway and energetics for methane-to-methanol conversion by first-row transition-metal oxide ions (MO^+s, where M is Sc, Ti, V, Cr, Mn, Fe, Co, Ni, and Cu). We introduce the electronic structures of MO^+ and the potential-energy diagrams for the methane hydroxylation by MO^+. In the FeO^+/CH_4 system, surface crossing seams were computed along the reaction pathway. Spin–orbit coupling (SOC) was calculated to estimate the probability of spin-inversion. SOC decreases along the reaction pathway, approaching zero in the product complex.

Keywords Density functional theory calculations · Methane activation · Spin states · Transition state · Surface crossing

1 Introduction

The development of catalysts for the selective oxidation of saturated hydrocarbons under mild conditions is a central research topic in modern chemistry [1–11]. Studies on the gas-phase reactions between bare transition-metal ions and hydrocarbons have provided a wealth of insight concerning the intrinsic interactions between the active site of catalysts and organic substrates [12–15]. The direct conversion of CH_4 to CH_3OH is thermodynamically more favorable than the commercial two-step process using synthesis gas (CO and H_2) [16]; it is therefore a current topic of interest in pure and applied chemistry.

The catalytic activity of bare transition-metal monoxide cations (MO^+) toward CH_4 is a key to the mechanistic aspects of direct CH_4 hydroxylation [17–30]. Schwarz and coworkers have systematically investigated the gas-phase reactions between the

Y. Shiota (✉) · K. Yoshizawa
Institute for Materials Chemistry and Engineering, Kyushu University, Motooka, Nishi-ku, Fukuoka 819-0395, Japan
e-mail: shiota@ms.ifoc.kyushu-u.ac.jp

K. Yoshizawa (ed.), *Direct Hydroxylation of Methane*,
https://doi.org/10.1007/978-981-15-6986-9_2

first-row MO^+ complexes and CH_4 [28, 30], demonstrating that late transition-metal MO^+ complexes can activate CH_4. The reaction efficiency and the CH_3OH branching ratio are strongly dependent on the metal. For example, FeO^+ efficiently reacts with CH_4 under ion cyclotron resonance (ICR) conditions, forming CH_3OH in 41% yield [21]. Although MnO^+ reacts with CH_4 very efficiently, the CH_3OH branching ratio is less than 1% [24]. CoO^+ exhibits low reactivity toward CH_4; however, the branching ratio to CH_3OH is 100% [27]. Both the reactivity and the CH_3OH branching ratio are high in the case of NiO^+ [28]. By contrast, the early transition-metal MO^+ complexes (ScO^+, TiO^+, and VO^+) exhibit no reactivity toward alkanes and alkenes because of their strong metal–oxo bonds. Interestingly, Sc^+ reacts with CH_3OH to yield ScO^+ and CH_4 in the gas phase [29], which is precisely the reverse reaction of CH_4 hydroxylation.

From density functional theory (DFT) computations, we determined the reaction pathway and its detailed energetics for the direct CH_4–CH_3OH conversion by FeO^+ [31–37]. Two transition states (TS1 and TS2) were confirmed from intrinsic reaction coordinate (IRC) and femtosecond dynamics calculations to correctly connect the reaction pathway. Of course, this reaction mechanism is not limited to the gas-phase CH_4 hydroxylation by bare MO^+ complexes. We have extended the reaction mechanism to the CH_4 hydroxylation catalyzed by soluble methane monooxygenase (sMMO) [37–42], particulate methane monooxygenase (pMMO) [43–49], and metal-exchanged zeolites [50–56].

2 Electronic Structures of MO^+ Ions

The gas-phase reaction between bare MO^+ cations and CH_4 is particularly interesting because it can be viewed as the simplest system for various oxidation reactions by enzymatic and zeolitic systems. Carter and Goddard [57] predicted the general bonding characters of the MO^+ complexes on the basis of all-electron ab initio generalized valence bond calculations. Moreover, Schwarz et al. used DFT computations to investigate the electronic structures of MO^+ complexes in an attempt to deduce the reactivity manifold of FeO^+, CoO^+, NiO^+, and CuO^+ with CH_4 [30, 58]. The reactivity of MO^+ can, in general, be understood by considering how the d-orbital occupation of the metal dictates the type of metal–oxygen bond formed, with early transition metals forming strong, unreactive triple bonds and late transition metals forming weak, reactive biradical double bonds. Figure 1 shows the orbital occupancies in the molecular orbitals of ScO^+ (d^0), FeO^+ (d^5), and CuO^+ (d^8), which can be partitioned into bonding (2σ and 1π), nonbonding (1σ and 1δ), and antibonding ($2\pi^*$ and $3\sigma^*$) orbitals. A metal–oxo bond and its catalytic function are strongly dependent on how the d orbitals are occupied.

In the $^1\Sigma^+$ ground state of ScO^+, all bonding orbitals are doubly occupied, suggesting a triple bond between the Sc and the O atoms. Therefore, the ground state of ScO^+ is analogous to that of dinitrogen, the dissociation energy of ScO^+ being 156.1 kcal/mol at the B3LYP level of theory. TiO^+ has a similar ScO^+ bond strength

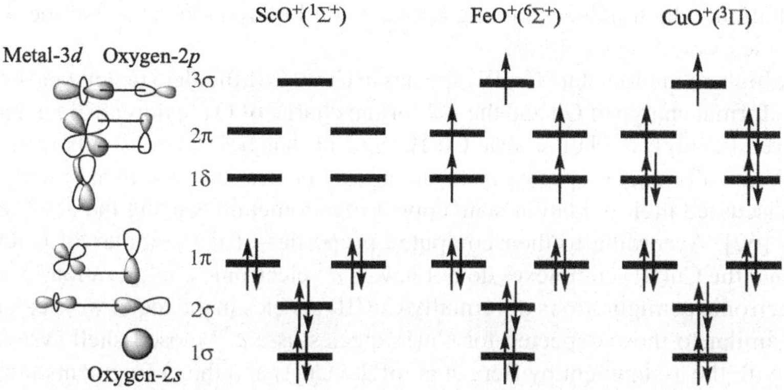

Fig. 1 Molecular orbitals of ScO^+, FeO^+, and CuO^+ and their occupancies

of 155.1 kcal/mol, and VO^+ has a slightly lower bond strength of 137.2 kcal/mol. The calculated bond dissociation energies (BDEs) are consistent with experimental values of 164 kcal/mol in ScO^+, 159 kcal/mol in TiO^+, and 134 kcal/mol in VO^+ [59]. The bonding natures of TiO^+ and VO^+ differ from that of ScO^+ because ScO^+ lacks an occupied *d* orbital. The singly occupied *d* orbital of TiO^+ and the doubly occupied *d* orbitals of VO^+ increase the interaction between the metal and the C–H bond of CH_4.

By contrast, FeO^+ and CuO^+, which have partially occupied antibonding orbitals, form a weak, reactive metal–oxo bond with a strong radical character. In the ground-state sextet of FeO^+, all of the bonding orbitals are doubly occupied and the unpaired electrons reside in the five 1δ, 2π, and 3σ orbitals; thus, its bonding nature resembles that of triplet dioxygen because of the singly occupied 2π antibonding orbital. The computed dissociation energy of FeO^+ of 75.2 kcal/mol is consistent with the reported experimental value of 81.2 kcal/mol [60]. The computed dissociation energy of 73.3 kcal/mol for the $^5\Delta$ state of CoO^+ is consistent with the reported experimental value of 77.2 kcal/mol [60]. In the $^4\Sigma^-$ ground state of NiO^+, all of the bonding orbitals and the 1δ nonbonding orbitals are doubly occupied and the unpaired electrons reside in the three 2π and 3σ orbitals. The computed dissociation energy of 69.3 kcal/mol for the $^4\Sigma^-$ state of NiO^+ is in good agreement with the reported experimental value of 63.3 kcal/mol [60]. The $^2\Sigma^+$ first excited state, which lies 11.6 kcal/mol above the ground state in energy, was formed from the $^4\Sigma^-$ ground state by a spin inversion within the 3σ orbital. According to B3LYP results for NiO^+, the $^2\Sigma^+$ excited state is energetically more stable than the $^2\Pi$ and $^2\Delta$ electronic configurations. The B3LYP energy for the 2Π state was 20.3 kcal/mol higher than that for the $^2\Sigma^+$ state. Because the low-lying sextet and quartet states have the same situation with respect to the occupied orbitals, the $^4\Sigma^- \rightarrow {}^2\Sigma^+$ transition for the bare NiO^+ should be a forbidden intersystem crossing according to El-Sayed's selection rules [61] for spin flip. The spin inversion at crossing seams requires spin–orbit perturbation, which induces mixing between the low-lying quartet and the doublet electronic excited states of NiO^+.

The high-oxidation-state Cu(III) species in CuO^+ is difficult to understand because the +3 formal charge of Cu and the −2 formal charge of O are unrealistic in general. However, the high-oxidation-state Cu(III) species has been increasingly recognized as a chemically useful species from unknown species in theoretical studies. Bera et al. discussed high-oxidation-state copper organometallics using the MP2 level of theory [62]. According to their computed properties of the complex of CuO^+ and ethylene, the Cu(III) complexes do not have a d^8 electronic configuration. Actually, the electronic configurations of formally Cu(III) species in the metal were calculated to be similar to those expected for Cu(I) species as a d^{10} closed-shell system. We agree with the assignment by Bera et al. of the Cu(I) and the O fragments for CuO^+ because the Mulliken charges were also calculated to be 1.26 for Cu and −0.26 for O at the B3LYP level of theory. The computed Cu–O bond energy in the $^3\Pi$ ground state was 37.6 kcal/mol, in good agreement with the reported experimental value of 37 kcal/mol [60].

Bridging these extremes in behavior are CrO^+ and MnO^+. The ground state of CrO^+ can oxidize saturated hydrocarbons larger than CH_4. The Cr–O bond strength for the ground $^4\Sigma^-$ state of CrO^+ is weak compared with that of other early transition-metal MO^+. The computed and experimental bond energies for Cr–O are 81.3 kcal/mol and 85.3 kcal/mol, respectively [60]. Carter and Goddard [57] suggested quintet ground states ($^5\Sigma^+$ or $^5\Pi$). In the case of the bond dissociation energy and the energetics of MnO^+, the difference between the $^5\Sigma^+$ or $^5\Pi$ states is negligible; thus, the exact energy separation remains unclear. According to computational results for MnO^+, the $^5\Sigma^+$ ground state of MnO^+ is lower in energy than the $^5\Pi$ state. The computed and experimental bond energies for Mn–O are 56.4 kcal/mol and 68 kcal/mol, respectively [60].

3 Potential-Energy Diagrams for the Methane-To-Methanol Conversion

As shown in Scheme 1, CH_4 hydroxylation by bare MO^+ ions is predicted to occur via a nonradical mechanism through heterolytic C–H bond cleavage [31]. In the first half of the reaction, a H-atom of the adsorbed CH_4 molecule in a reactant complex (**RC**) migrates to the O atom of the active site via a four-centered transition state (TS1), leading to the formation of a hydroxo intermediate (**HI**), where methyl and hydroxo moieties are formed on the active site. In the second half of the reaction, HO–CH_3 recombination occurs via a three-centered transition state (TS2) to form a product complex (**PC**).

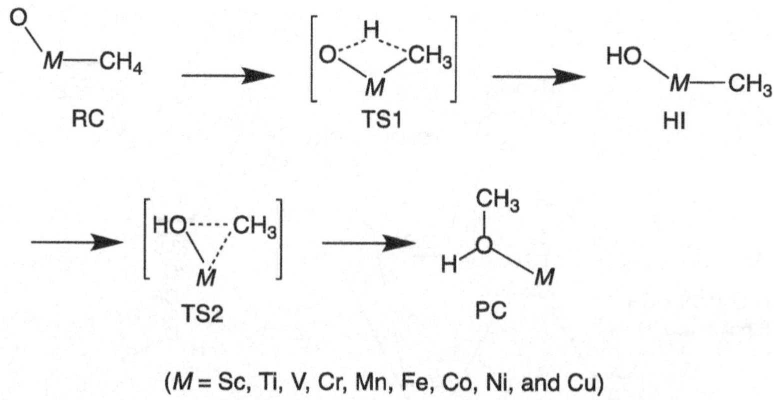

Scheme 1 Potential-energy diagrams for the methane-to-methanol conversion

3.1 Conversion of Methane to Methanol by ScO⁺, TiO⁺, and VO⁺

Figure 2 shows the energy diagrams for the CH_4–to–CH_3OH conversion by ScO^+, TiO^+, and VO^+ along the two-step concerted reaction pathway. One possible restriction for this reaction is that the formation of the ground-state products [$Sc^+(^3D)$, $Ti^+(^4F)$, and $V^+(^5D)$ + CH_3OH] from the ground-state reactants [$ScO^+(^1\Sigma^+)$, $TiO^+(^2\Delta)$, and $VO^+(^3\Sigma^-)$ + CH_4] is spin-forbidden because surface crossing between the high-spin and low-spin states occurs in the course of the reaction. Experiments carried out in the groups of Schwarz and Armentrout have shown that early transition-metal oxide ions hardly react with CH_4, the reaction efficiencies being less than 0.01% [28].

In the singlet ground state, ScO^+ interacts with CH_4 with a binding energy of 13.5 kcal/mol. The OSc^+–CH_4 species can be formally viewed as a d^0 system; thus, there is no net interaction between the *d*-block orbitals of ScO^+ and the occupied orbitals of CH_4. The computed binding energies for OTi^+–CH_4 and OV^+–CH_4 are 15.4 kcal/mol and 16.8 kcal/mol, respectively. Because the singly occupied *d* orbital of TiO^+ and the doubly occupied *d* orbitals of VO^+ contribute to the increase in interaction between the metal and the CH_4, the binding energy of **RC** increases from Sc to V.

The bound CH_4 in the reactant complex undergoes a concerted 1,3-hydrogen migration, leading to **HI**. In the ground state of ScO^+, the potential energy for TS1 lies above the dissociation limit by 12.3 kcal/mol. The potential energy for the TS1 of VO^+ was calculated to be 17.6 kcal/mol higher in energy than that calculated for TiO^+ (16.0 kcal/mol). The singlet **HI** corresponding to CH_3–Sc^+–OH was calculated to be −20.7 kcal/mol, as measured from the dissociation limit on the singlet state. On the doublet state for the TiO^+/CH_4 system and the triplet state for the VO^+/CH_4 system, **HI** lies below the dissociation limit by −14.2 kcal/mol and −13.2 kcal/mol, respectively.

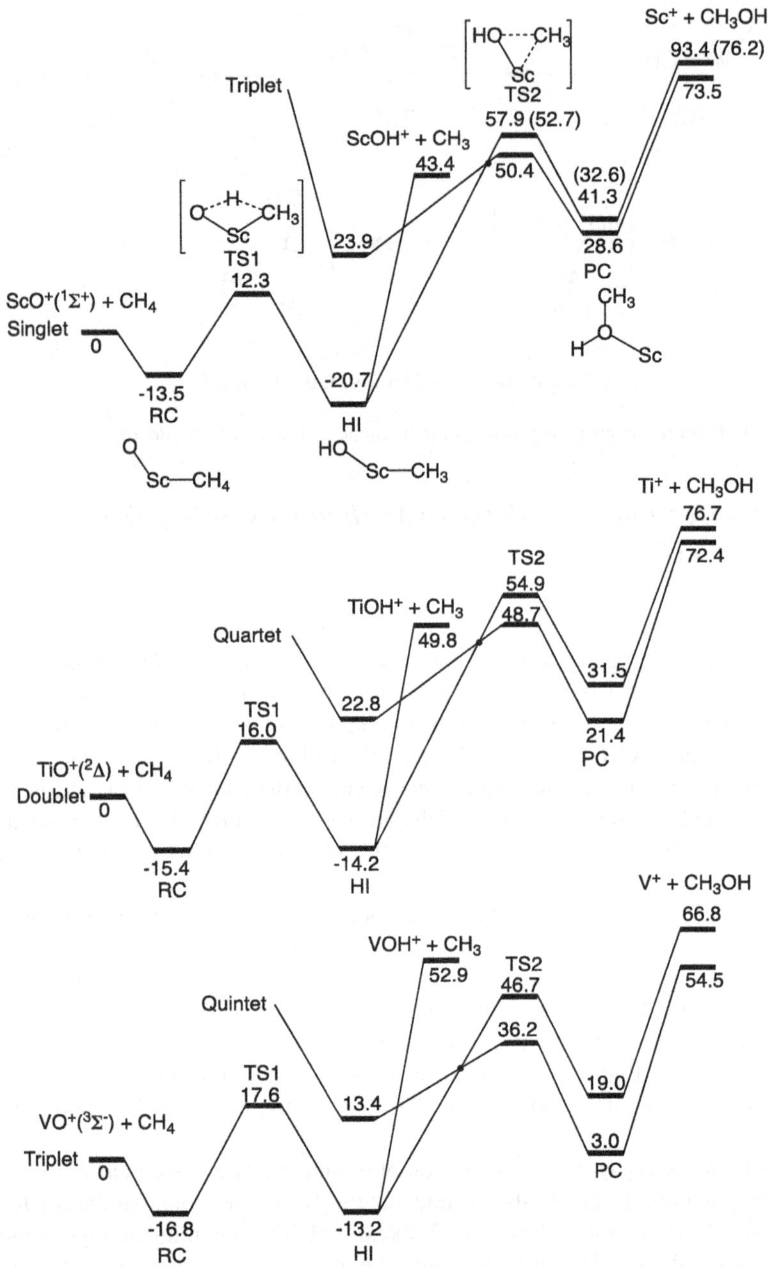

Fig. 2 Potential-energy diagrams (including the zero-point energy) along the reaction pathway for $MO^+ + CH_4 \rightarrow M^+ + CH_3OH$, in the high-spin and low-spin states. Relative energies are in kcal/mol. Reproduced from Ref. [36]. Copyright 2000 American Chemical Society

The formation of the **PC** corresponding to $M^{+}(CH_3OH)$ occurs in a concerted manner via 1,2-methyl migration on **HI**. The potential-energy surface of the singlet state in the ScO^{+}/CH_4 system lies below that of the triplet state in the ScO^{+}/CH_4 system, and their relative order is reversed in TS2, as shown in Fig. 2. Thus, there should be a crossing of the triplet and singlet potential-energy surfaces that occurs near the region where TS2 is formed. Similar crossings should also occur in the TiO^{+}/CH_4 and VO^{+}/CH_4 systems.

In this section, we consider the reverse reaction: $M^{+} + CH_3OH \rightarrow MO^{+} + CH_4$. Irigoras, Fowler, and Ugalde [33] carried out ab initio calculations for the ground state of the analogous reverse process, MO^{+} + H – H, formed by the reaction of $M^{+} + H_2O$ in the gas phase. The potential-energy surfaces indicate that the low-spin first excited state should provide high reactivity because of the low-spin surface leading to $MO^{+} + CH_4$ without a spin-forbidden surface crossing. The energetics of the early transition-metal MO^{+} complexes indicate that the reverse reactions are preferred over the corresponding forward reactions. These calculations explain an experimental observation that the early transition metals prefer an oxygen acceptor over an oxygen donor.

3.2 *Conversion of Methane to Methanol by CrO^{+} and MnO^{+}*

Figure 3 shows computed potential-energy diagrams along the entire reaction pathway, $CrO^{+} + CH_4 \rightarrow Cr^{+} + CH_3OH$, in the quartet and doublet states. The chemistry of CrO^{+} and MnO^{+} complexes shows the bridging between the early and the late transition-metal MO^{+} complexes. Kang and Beauchamp [18, 19] carried out the gas-phase reactions of CrO^{+} with alkanes and found that CrO^{+} can convert ethane to ethanol without forming a byproduct. Because the high-spin sextet potential-energy surface lies above the low-spin quartet potential-energy surface, a crossing between the sextet and quartet surfaces occurs only once near TS2. In the quartet ground state, CrO^{+} reacts with CH_4, and the binding energy for the **RC**, OCr^{+}–(CH_4), was calculated to be 21.5 kcal/mol. TS1 in the CrO^{+}/CH_4 system differs from those in the other early transition-metal MO^{+}/CH_4 systems (e.g., ScO^{+}, TiO^{+}, and VO^{+}/CH_4) because TS1 lies at a lower energy that the dissociation limit $CrO^{+} + CH_4$ by 0.7 kcal/mol. Because the initial energy is sufficient for passing over TS1, the reaction efficiency of CrO^{+} for the H-atom abstraction should be substantial compared with those of ScO^{+}, TiO^{+}, and VO^{+}. The quartet **HI**, CH_3–Cr^{+}–OH, was calculated to be −34.2 kcal/mol, as measured from the dissociation limit.

In TS2, the potential-energy surface of the quartet state lies below that of the sextet state and their relative positions are reversed. TS2 of the sextet spin state is more stable than that of the quartet spin state, and the stabilization energy is 16.7 kcal/mol because of a spin inversion from the quartet state to the sextet state. The spin inversion contributes to the reaction via TS2. TS2 in the CrO^{+}/CH_4 system was calculated to be −6.3 kcal/mol in the sextet spin sate, whereas the product complex of the sextet spin state was calculated to be −45.3 kcal/mol. The spin-conserving process in

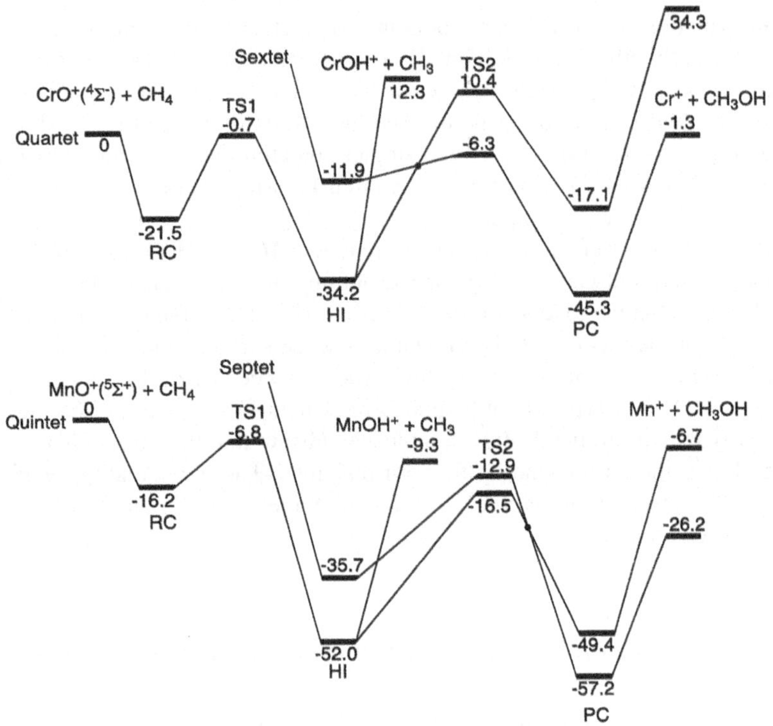

Fig. 3 Potential-energy diagrams (including the zero-point energy) along the reaction pathway, $MO^+ + CH_4 \rightarrow M^+ + CH_3OH$, in the high-spin and low-spin states. Relative energies are in kcal/mol. Reproduced from Ref. [36]. Copyright 2000 American Chemical Society

the quartet state is uphill to Cr^+(quartet) + CH_3OH by 34.3 kcal/mol, whereas the potential energy of Cr^+(sextet) + CH_3OH is almost isoenergetic with the dissociation limit. This exothermicity is in good agreement with the value of 4 kcal/mol estimated from ion-beam experiments [26].

Schwarz's group reported that MnO^+ reacts with CH_4 in the quintet ground state [24]. For the reaction of MnO^+ with CH_4, the H-atom abstraction reaction to form $MnOH^+$ is a main process and the conversion of CH_4 to CH_3OH mediated by MnO^+ is observed as a minor process. Figure 4 shows computed potential-energy diagrams along the reaction pathway, $MnO^+ + CH_4 \rightarrow Mn^+ + CH_3OH$, in the quintet and septet states. The general features of MnO^+/CH_4 systems are very smaller to those of the potential energy for the CrO^+ case. In the **RC**, the binding energy of a CH_4 molecule to MnO^+ is 16.2 kcal/mol, and the relative energy of TS1 in the MnO^+/CH_4 system is −6.8 kcal/mol. This value is smaller than −0.7 kcal/mol for CrO^+ and about 15 kcal/mol for early transition-metal MO^+ complexes.

The relative energies of TS2 are −12.9 kcal/mol in the quintet state and −16.5 kcal/mol in the septet, as measured from the dissociation limit. The energy splitting of TS2 between the quartet and septet potential-energy surfaces decreases to

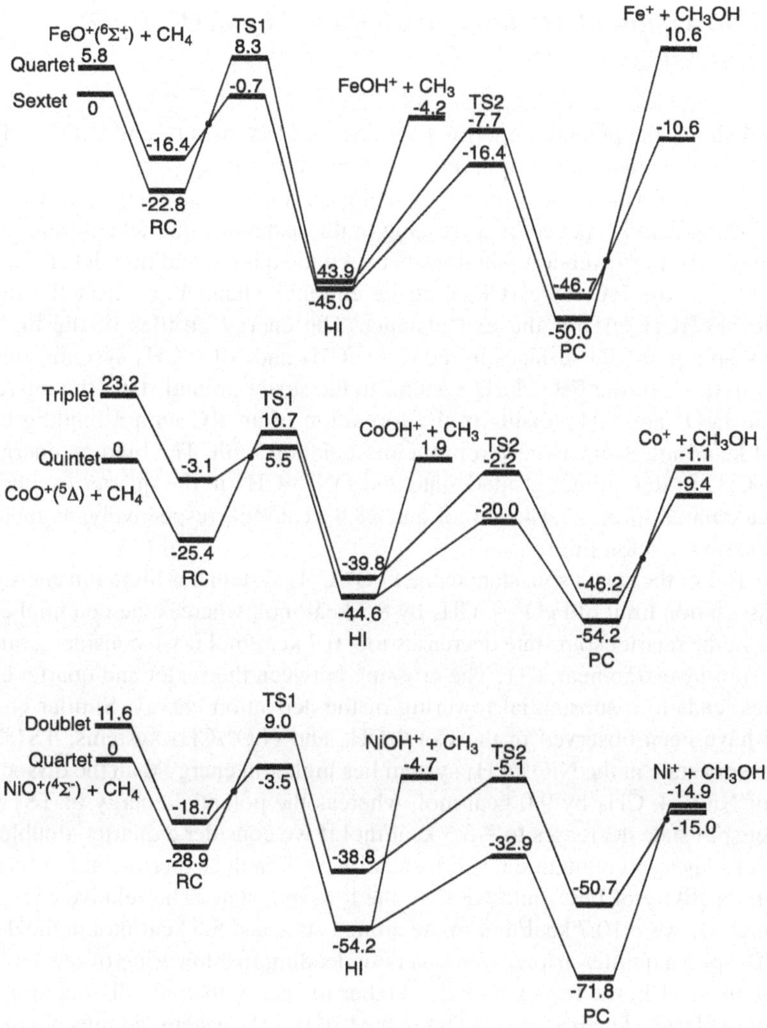

Fig. 4 Potential-energy diagrams (including the zero-point energy) along the reaction pathway, $MO^+ + CH_4 \rightarrow M^+ + CH_3OH$, in the high-spin and low-spin states. Relative energies are in kcal/mol. Reproduced from Ref. [36]. Copyright 2000 American Chemical Society

3.6 kcal/mol from 12.3 kcal/mol of **HI**. The surface crossing between the quartet and septet energy surfaces occurs in the vicinity of TS2. Because all reaction species in the MnO^+/CH_4 system lie lower in energy than the dissociation limit, the initial energy is sufficient to overcome TS1 and TS2. Therefore, the crossing point of the MnO^+/CH_4 system does not contribute to the reaction via TS2 compared with the crossing point of the CrO^+/CH_4 system. The overall reaction energies in the MnO^+/CH_4 system are exothermic: 6.3 kcal/mol in the quintet state and 26.2 kcal/mol in the septet state.

3.3 Conversion of Methane to Methanol by FeO^+, CoO^+, and NiO^+

Figure 4 shows the potential-energy diagrams for FeO^+, CoO^+, and NiO^+ with CH_4 along the entire reaction pathway. FeO^+ exhibits close-lying sextet and quartet spin state potential energies that differ by 5.8 kcal/mol. A low-lying energy pathway in the FeO^+/CH_4 system is opened by a crossing of the sextet and quartet potential-energy surfaces. Two spin-inversion junctions between the quartet and doublet surfaces are involved near the **RC**, $OFe^+(CH_4)$, at the entrance channel and near the product complex $Fe^+(CH_3OH)$ at the exit channel. The energy profiles of the high-spin and low-spin potential surfaces in the CoO^+/CH_4 and NiO^+/CH_4 systems are quite similar to those of the FeO^+/CH_4 system. In the sextet ground state, the interaction between FeO^+ and CH_4 results in the formation of an **RC** with a binding energy of 25.4 kcal/mol, as measured from the dissociation limit. The binding energies of $OCo^+–CH_4$ in the quintet ground state and $ONi^+–CH_4$ in the quartet ground state were calculated to be 25.4 kcal/mol and 28.9 kcal/mol, respectively, as measured from the dissociation limit.

The TS1 of the sextet spin state in the FeO^+/CH_4 system lies higher in energy than the dissociation limit of $FeO^+ + CH_4$ by 8.3 kcal/mol, whereas the potential energy of TS1 in the quartet spin state decreases to −0.7 kcal/mol if we consider a quartet–sextet spin inversion near TS1. The crossing between the sextet and quartet energy surfaces leads to a substantial lowering of the activation energy. Similar crossing points have been observed in the CoO^+/CH_4 and NiO^+/CH_4 systems. TS1 of the quartet spin state in the NiO^+/CH_4 system lies higher in energy than the dissociation limit of $NiO^+ + CH_4$ by 9.0 kcal/mol, whereas the potential energy of TS1 in the doublet spin state decreases to −3.5 kcal/mol if we consider a quartet–doublet spin inversion. These computational results readily explain the experimental observation of high reactivity of FeO^+ and NiO^+ in the low-spin state. The relative energies of TS1 in CoO^+ were 10.7 kcal/mol in the quintet state and 5.5 kcal/mol in the doublet state. Despite a quintet–triplet spin inversion leading to a lowering of the activation energy, the TS1 in the triplet state lies higher in energy than the dissociation limit. Therefore, the C–H activation via TS1 in the CoO^+/CH_4 system requires not only the spin-inversion process but also additional energy. The **HI**s, $HO–Fe^+–CH_3$, $HO–Co^+–CH_3$, and $HO–Ni^+–CH_3$ are energetically more stable than with the corresponding **RC**s by ~20 kcal/mol. The relative energies of TS2 in the FeO^+/CH_4 system are −7.7 kcal/mol in the sextet state and −16.4 kcal/mol in the quartet state. The relative energies of TS2 in the CoO^+/CH_4 and NiO^+/CH_4 systems are −20.0 kcal/mol in the triplet state and −32.9 kcal/mol in the doublet state. Because the potential energies of TS2 are lower than those of TS1 in the FeO^+/CH_4, CoO^+/CH_4, and NiO^+/CH_4 systems, the rate-determining step in the two-step concerted mechanism is not the second-step reaction with the methyl migration but the first-step reaction that includes C–H bond cleavage.

For a comparison of the reactions of MnO^+, FeO^+, CoO^+, and NiO^+ with CH_4 on the basis of experimental data [28], the reaction efficiencies ϕ and the product

Table 1 Measured reaction efficiencies (ϕ) and product branching ratios for the reaction of MO^+ with CH_4 [28]

MO^+	ϕ	$MOH^+ + CH_3$	$MCH_2^+ + H_2O$	$M^+ + CH_3OH$
MnO^+	40	100	–	<1
FeO^+	20	87	2	41
CoO^+	0.5	–	–	100
NiO^+	20	–	–	100

branching ratios for the reaction of MO^+ with CH_4 are listed in Table 1. The observed reaction efficiencies of MnO^+, FeO^+, NiO^+, and CoO^+ toward CH_4 are 40, 20, 0.5, and 20%, respectively. For example, MnO^+ reacts in a fast process by H-abstraction to yield mainly $MnOH^+$. By contrast, the reaction efficiency of CoO^+ toward CH_4 is very low and the branching ratio to CH_3OH is 100%. As previously stated, the rate-determining step of this reaction for the late transition-metal MO^+/CH_4 systems is the first-step reaction via TS1. The relative energy of TS1 in the MnO^+/CH_4 system is lower than those in the FeO^+/CH_4 and NiO^+/CH_4 systems; in addition, this value is much smaller than that in the CoO^+/CH_4 system. The CH_3OH branching ratios are 100% in CoO^+ and NiO^+, 41% in FeO^+, and <1% in MnO^+. Methyl-radical formation by M–C bond dissociation, $HO\text{–}M^+\text{–}CH_3 \rightarrow MOH^+ + {}^{\bullet}CH_3$, is the competitive reaction, which gives a byproduct for the CH_4-to-CH_3OH conversion. The energy for $NiOH^+ + {}^{\bullet}CH_3$ lies 28.2 kcal/mol above TS2 of the doublet state that leads to the **PC**, and the energy for $CoOH^+ + {}^{\bullet}CH_3$ lies 21.9 kcal/mol above TS2 of the triplet state. Because DFT calculations suggest that the methyl radicals formed in the CoO^+/CH_4 and NiO^+/CH_4 systems lie higher in energy than TS2 corresponding to the formation of CH_3OH, the CH_3OH-branching pathway in the CoO^+/CH_4 and NiO^+/CH_4 systems is likely to be energetically preferred according to the potential-energy profile for the conversion of CH_4 to CH_3OH. By contrast, the methyl-radical branching ratio is 100% in MnO^+ and 41% in FeO^+. The energy for $MnOH^+ + {}^{\bullet}CH_3$ lies only 7.2 kcal/mol above TS2 in the quintet state, and that for $FeOH^+ + {}^{\bullet}CH_3$ lies 12.2 kcal/mol above TS2 in the quartet state. Because the CH_3OH branching ratio has been rationalized from the energy splitting between the reaction pathway to radical formation ($MOH^+ + {}^{\bullet}CH_3$) and to CH_3OH formation via TS2, the CH_3OH branching ratios of the late transition-metal MO^+/CH_4 system can be explained in a similar manner.

3.4 *Conversion of Methane to Methanol by CuO^+*

We here examine the potential-energy diagram along the entire reaction pathway, $CuO^+ + CH_4 \rightarrow Cu^+ + CH_3OH$, in the triplet and singlet states. The triplet energy of TS1 was calculated to be −13.0 kcal/mol, as measured from the dissociation limit. A transition state is found using the unrestricted B3LYP method, and the open-shell

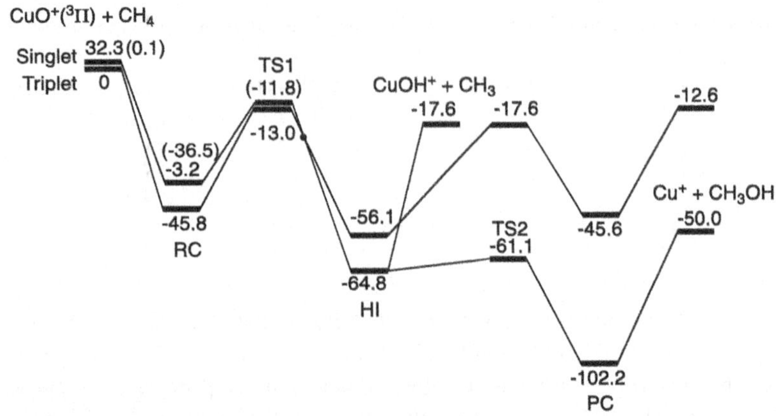

Fig. 5 Potential-energy diagrams (including the zero-point energy) along the reaction pathway, $CuO^+ + CH_4 \rightarrow Cu^+ + CH_3OH$, in the high-spin and low-spin states. Relative energies are in kcal/mol. Values in parentheses are energies in the open-shell singlet. Reproduced from Ref. [36]. Copyright 2000 American Chemical Society

singlet TS1 lies higher in energy than the triplet TS1 by 1.2 kcal/mol. Therefore, the surface crossing between the triplet and the singlet energy surfaces occurs in the vicinity of **HI**, in which the energy splitting is 8.7 kcal/mol. The relative energies of TS2 were calculated to be −61.1 kcal/mol in the singlet state and −17.6 kcal/mol in the triplet state. The reaction pathway via TS2 in the singlet state is energetically more favorable than that leading to $CuOH^+$ and CH_3. The activation barrier relative to **HI** was calculated to be only 3.7 kcal/mol, and the potential energy for **PC** in the singlet state is very low-lying at −102.2 kcal/mol. This reaction is therefore preferred from the viewpoint of energetics. The overall reaction is 50.0 kcal/mol exothermic, with spin inversion from the triplet state to the singlet state. We thus conclude that the CH_4–to–CH_3OH conversion should be best mediated by CuO^+. The DFT calculations suggest that CuO^+ should exhibit very high activity toward CH_4 oxidation. In 2011, a theoretical prediction was reported for the reaction between CuO^+ and CH_4 under ICR conditions [30] (Fig. 5).

4 Surface Crossing and Spin–Orbit Coupling

A topic of interest in the CH_4–to–CH_3OH conversion by metal–oxo species is the electronic process of spin inversion, which can occur in the vicinity of a crossing region of two potential-energy surfaces of different spin states. In contrast to organic reactions, which proceed on a single potential-energy surface in most cases, reactions mediated by organometallic systems can proceed on more than one potential-energy surface. For example, the reactions of $FeO^+$with H_2 [63, 64] and CH_4 [36, 37] have been analyzed in terms of the corresponding potential-energy surfaces. In this reaction, the crossing of the quartet and sextet potential-energy surfaces occurs twice: once in the entrance channel, and once in the exit channel. The extent of spin–orbit coupling (SOC) is relatively large in the entrance channel, and it decreases along the oxidation process, approaching zero in the exit channel. The overall reaction is limited by SOC effects at spin-inversion points. It turns out that, in addition to the classical factors such as the potential barrier heights, the surface crossing and the SOC factor play important roles because a crossing between the high-spin and the low-spin states constitutes a distinct mechanistic step along the reaction coordinate. This phenomenon, known as "two-state reactivity," is likely to be a key feature in organometallic chemistry [65].

4.1 Crossing Seams of Potential-Energy Surfaces

Crossing seams and spin-inversion processes are discussed in this subsection. The B3LYP functional predicts a lowering of the activation energy to 22.1 kcal/mol through spin inversion. As previously discussed, the transition is expected to occur from the ground sextet state to the lowest quartet state in the region prior to TS1. To obtain more detailed information about the potential-energy surfaces, we performed single-point computations of the sextet state as a function of the structural change along the intrinsic reaction coordinate (IRC) of the quartet state and vice versa; such analyses will indicate the energy-minimum and energy-maximum crossing points [34]. The crossing seam of high- and low-spin potential-energy surfaces satisfies the constraint of high-spin energy = low-spin energy. Because the crossing seam shows the line as a ridge between potential-energy surfaces, a crossing seam plays an important role in the nonadiabatic transition between the potential-energy surfaces. The energy profiles for the high- and low-spin states are displayed in Figs. 6 and 7. One crossing-seam point (**SI**) is on the quartet IRC path, and three crossing-seam points (**SII**, **SIII**, and **SIV**) are on the sextet IRC path.

The solid and the dotted lines in Fig. 6 indicate the computed potential-energy profiles of the quartet and the sextet states, respectively, as a function of the structural change along the quartet IRC. The relative energies are measured from the total energy of the reactant complex of the sextet state as a standard. The activation energy for TS1 on the sextet surface is 38.5 kcal/mol, and the exothermicity is ~20.0 kcal/mol. The potential energy for the TS1 on the quartet surface is lower than the sextet potential energy by 13.9 kcal/mol. As shown in Fig. 6a, the first crossing-seam point, **SI**, is observed at $s = -2.80$ with an energy of 13.9 kcal/mol. The quartet energy surface begins to cross the sextet surface from **SI** because the sextet energy on the sextet IRC is still more stable than the quartet energy on the quartet IRC. Once **SI** has been reached, the quartet IRC continues a stable pathway toward **PC** and the energy gap is ~15 kcal/mol in the region between TS1 and TS2. The relative energies for TS2 in the quartet and sextet states are 5.93 kcal/mol and 21.8 kcal/mol, respectively. Because the relative energy for TS2 in the sextet spin

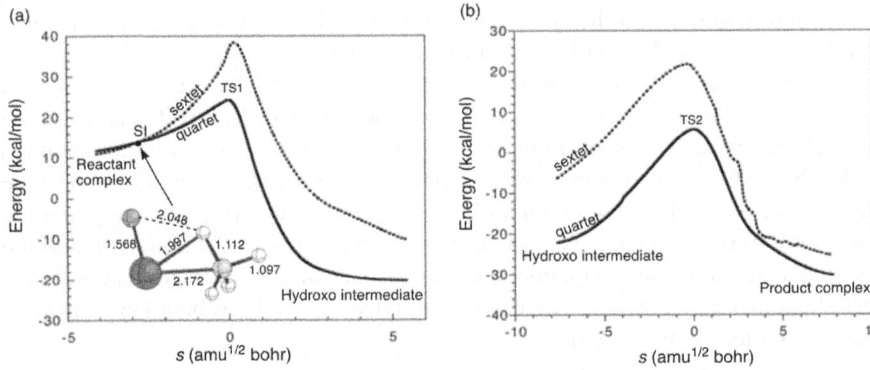

Fig. 6 Potential energies along the quartet IRC **a** from the reactant complex to the hydroxo intermediate and **b** from the hydroxo intermediate to the product complex. Reproduced from Ref. [34]. Copyright 1999 American Institute of Physics

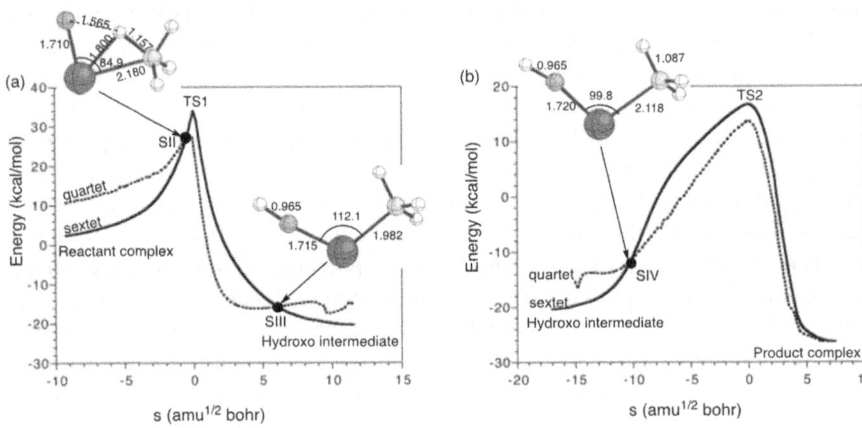

Fig. 7 Potential energies along the sextet IRC **a** from the reactant complex to the hydroxo intermediate and **b** from the hydroxo intermediate to the product complex. Reproduced from Ref. [34]. Copyright 1999 American Institute of Physics

state is more unstable than the potential energy for the dissociation step of an Fe–C bond, the reaction step proceeding to TS2 should occur in the quartet spin state.

In Fig. 7, the solid and the dotted lines indicate computed potential-energy profiles of the sextet and the quartet states, respectively, as a function of structural change along the sextet IRC. The transition to the quartet state from the sextet state stabilizes the potential energies of both TS1 and TS2 by only ~5 kcal/mol. Thus, the quartet potential-energy surface along the sextet IRC does not substantially improve the energetics. These results are consistent with the prediction based on the theory of IRC, which is regarded as an optimized reaction pathway. Because the IRC can stabilize only the potential energy for the same spin state, the energy gap between the sextet and quartet spin states tends to be enlarged for the quartet IRC and to be reduced for the sextet IRC in the region where the quartet surface is a low-energy path.

SII is found in the vicinity of TS1 at $s = -0.50$ with an energy of 27.3 kcal/mol and is only 0.05 kcal/mol lower in energy than TS1 on the quartet surface. **SII** exhibits C_s symmetry, a C–O distance of 1.565, and one long C–H distance of 1.157 Å. After **SII**, the sextet surface level shifts over the quartet surface without the intersection. Of course, the activation energy for TS1 is important as a spin-inversion point for controlling the reaction efficiency. In the overview of this reaction step, the Fe–O distance controls the energy gap through the sextet IRC or the quartet IRC as a parallel for the reaction coordinate. For instance, when an FeO^+ molecule with an Fe–O distance of 1.700 Å collides with a CH_4 molecule, the two potential-energy surfaces along the reaction coordinate behave like that in the sextet IRC. The difference between the sextet and quartet IRC results is due to the difference in the Fe–O distance at the entrance channel, and we emphasize that the potential energy in the sextet state changes dramatically. That is, the quartet potential-energy surface is unaffected by the Fe–O distance.

In Fig. 7a, the quartet potential-energy surface involves a local minimum point at $s = 4.6$ with an energy of −16.1 kcal/mol; a slight increase is observed after this point. The sextet potential energy continues to decrease until the product is formed, and **SIII** is found at $s = 6.0$ with an energy of −15.5 kcal/mol. **SIV** is found with an energy of −10.3 kcal/mol, as shown in Fig. 8b. **SIII** and **SIV** are the local maximum points on the seam and are located at both ends of the seam line. In the region of the lowest crossing seam near the hydroxy intermediate, the minimum-energy point on the seam indicates the transition state for the spin inversion from one hydroxy

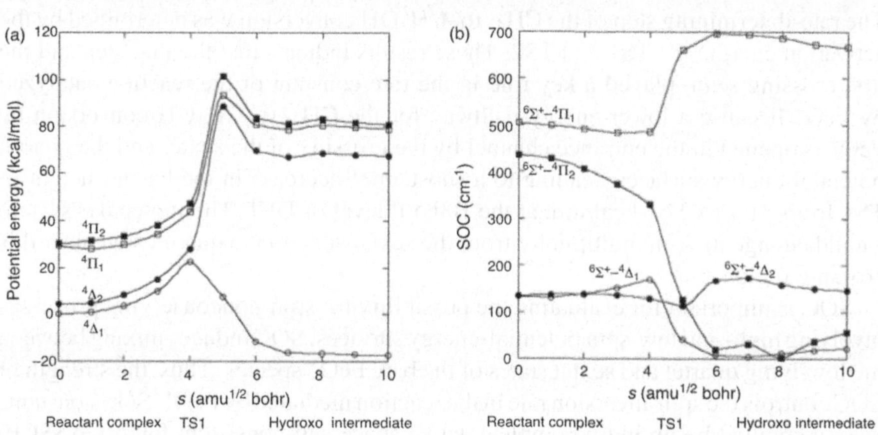

Fig. 8 **a** Potential energies at the CASSCF level and **b** SOC values along the quartet IRC. The quartet IRC connects the reaction pathway from the reactant complex ($s = 0$) to the hydroxo intermediate ($s = 10$) via TS1 ($s = 4.0$). Reproduced from Ref. [37]. Copyright 2003 American Institute of Physics

intermediate to another hydroxy intermediate; we can estimate that the activation energy for this transition state is less than 10 kcal/mol.

We summarize some features of the sextet and the quartet potential-energy surfaces along the reaction pathway for the CH_4–to–CH_3OH conversion. The ground state of FeO^+ is the sextet state; thus, that of the reactant complex is also the sextet state because the interaction between CH_4 and FeO^+ is not strong in the reactant complex. The potential-energy surface of the sextet state begins to increase from the reactant complex and contacts with the quartet IRC valley at point **SI**. The sextet potential-energy surface still continues in increase while maintaining contact with the quartet potential-energy surface; finally, the sextet IRC valley contacts the quartet potential-energy surface at point **SII**. After passing point **SII**, the sextet potential-energy surface completely lies above the quartet potential-energy surface. Consequently, there is a crossing seam of the two potential-energy surfaces in the region between **SI** and **SII**, where **SI** and **SII** are the energy-minimum and energy-maximum crossing points, respectively, in the first crossing seam. After the IRCs of the sextet and the quartet states, the molecular structures become different after passing TS1; as a result, the second crossing seam appears at point **SIII** and disappears at point **SIV**. The two potential-energy surfaces are separated again after passing point **SIV**, with the sextet potential energy lying slightly above the quartet potential energy until the **PC** is formed. The third crossing seam exists in the exit channel, and it plays a dominant role in the elimination of product CH_3OH.

4.2 *Spin–Orbit Coupling of Methane Conversion*

The rate-determining step of the CH_4–to–CH_3OH conversion was determined by the activation energies of TS1 and TS2. These results indicate that the first step and the first crossing seam played a key role in the rate constant of the reaction catalyzed by FeO^+ because a lower-energy pathway for the CH_4–to–CH_3OH conversion by FeO^+ is opened in the entrance channel by the crossing of the sextet and the quartet potential-energy surfaces, leading to a substantial decrease in the barrier height of TS1 from 31.1 to 22.1 kcal/mol at the B3LYP level of DFT. The molecular system should change its spin multiplicity from the sextet state to the quartet state near this crossing region.

SOC is important for evaluating the possibility for spin-nonconserving processes involving high- and low-spin potential-energy surfaces. SOC induces mixing between the low-lying quartet and sextet states of the bare FeO^+ species. Thus, the strength of SOC controls the spin-inversion rate in the reaction mediated by FeO^+. SOC constants were computed by ab initio complete active space self-consistent field (CASSCF) theory along the reaction pathway of the oxidation process of CH_4, $CH_4 + FeO^+ \rightarrow CH_3OH + Fe^+$ [37]. The process involves three spin crossing seams of the quartet and sextet potential-energy surfaces. This electronic feature is the most important aspect in this reaction pathway in that the FeO^+/CH_4 system should change its spin multiplicity from the sextet to the quartet state and that this effect should lead to a substantial decrease in the barrier heights of the transition states. An intersystem crossing is expected to occur in the transition state for the C–H bond breaking by the iron–oxo species.

Figure 8 shows computed potential-energy surfaces in the quartet state at the CASSCF level and SOC values between the quartet and sextet states as a function of reaction coordinate s. The reaction coordinate s locates the reactant complex at $s = 0$, TS1 at $s = 4.0$, and the hydroxo intermediate at $s = 10.0$. According to the B3LYP results, the energy gap between the low-lying quartet and ground sextet states decreases to zero in the first crossing seam located near $s = 1.3$. The four quartet states ($^4\Delta_1$, $^4\Delta_2$, $^4\Pi_1$, and $^4\Pi_2$) exhibit a gradual and smooth change in potential energies, as indicated in Fig. 8a. The relative energies are measured from the total energy of the $^4\Delta_1$ state in the reactant complex as a standard. In the low-lying $^4\Delta_1$ state, the activation energy is 22.6 kcal/mol and this reaction step is 17.3 kcal/mol exothermic. The potential energies of the $^4\Delta_1$ and $^4\Delta_2$ states are close-lying prior to TS1. After the reaction passes through TS1, degenerate energies of the $^4\Delta_1$ and $^4\Delta_2$ states are separated. The $^4\Pi_1$ and $^4\Pi_2$ states in the reactant complex lie approximately 30 kcal/mol above the $^4\Delta_1$ state. After all, the $^4\Delta_2$, $^4\Pi_1$, and $^4\Pi_2$ states lie much higher than the $^4\Delta_1$ state in the hydroxo intermediate. Thus, the $^4\Delta_2$, $^4\Pi_1$, and $^4\Pi_2$ states can be reasonably neglected in the spin inversion after TS1.

The SOC values between the four quartet states and the sextet state are indicated in Fig. 8b. The $^6\Sigma^+ - {}^4\Delta$ SOC is a key to understanding the spin-forbidden transition that occurs in the vicinity of the first crossing seam in which the spin multiplicity of the ground state is changed from the sextet state to the quartet state. Starting from 133 cm^{-1} at $s = 0$, the $^6\Sigma^+ - {}^4\Delta$ SOC values increase to 169.2 cm^{-1} at $s = 4.0$. The strength of the $^6\Sigma^+ - {}^4\Delta_1$ SOC increases and the energy gap between the $^6\Sigma^+$ and $^4\Delta_1$ states decreases as the reaction coordinate progresses. Thus, the $^6\Sigma^+ - {}^4\Delta_1$ SOC is particularly effective in the spin transition. The $^6\Sigma^+ - {}^4\Delta_1$ SOC values start from 133 cm^{-1} at $s = 0$, resulting in 128.4 cm^{-1} at $s = 4.0$. This value is also sufficiently large to contribute to the spin transition. Therefore, we expect that the $^6\Sigma^+ - {}^4\Delta_1$ SOC and $^6\Sigma^+ - {}^4\Delta_2$ SOC transitions should occur near the first crossing seam. The $^6\Sigma^+ - {}^4\Pi_1$ SOC and $^6\Sigma^+ - {}^4\Pi_2$ SOC values of the reactant complex are 500 and 430 cm^{-1}, respectively. However, the $^6\Sigma^+ - {}^4\Pi$ transition is unlikely to contribute to the low-lying reaction pathway through the spin inversion because the crossing seam between the $^6\Sigma^+$ and $^4\Pi$ states lies above TS1 of the $^4\Delta_1$ state. After C–H bond dissociation occurs in TS1, the $^4\Delta_2$ state lies 80 kcal/mol above the $^4\Delta_1$ state. Therefore, only the $^6\Sigma^+ - {}^4\Delta_1$ SOC substantially contributes to the spin transition once the reacting system passes through TS1. The $^6\Sigma^+ - {}^4\Delta_1$ SOC value starts with 40.3 cm^{-1} at $s = 5.0$ and continues to decrease to 4.5 cm^{-1} at $s =$ 7.0. After this minimum point, the $^6\Sigma^+ - {}^4\Delta_1$ SOC value increases to 21.4 cm^{-1} in the hydroxo intermediate. The $^6\Sigma^+ - {}^4\Delta_1$ SOC is relatively small after TS1. Our analysis predicts that the spin inversion between the $^6\Sigma^+$ and $^4\Delta$ states should occur in the vicinity of the first crossing seam ($s = 1.3$), whereas the spin inversion hardly occurs after TS1 ($s = 4.0$).

The potential energy of the $^4\Delta_1$ state decreases after the system passes through TS1, and the second crossing seam occurs in the vicinity of the hydroxo intermediate.

This intersection arises from the fact that the equilibrium geometries of the hydroxo intermediates of the spin states differ, whereas the quartet state and the sextet state are close in energy. Because the $^6\Sigma^+ - {}^4\Delta_1$ SOC value is 21.4 cm^{-1} in the hydroxo intermediate, the transition between the $^6\Sigma^+$ and $^4\Delta_1$ states is unlikely to occur in the second crossing seam.

The second half of the reaction pathway is responsible for the formation of the product complex via TS2. The essential process in the second half of the reaction is the dissociation of the Fe–C bond and the formation of the C–O bond. We calculated the potential-energy surfaces and the SOC values in the second half of the reaction pathway, as shown in Fig. 9. The reaction proceeds from the hydroxo intermediate ($s = 0$) to the product complex ($s = 16$) via TS2 ($s = 8.0$). Here, the relative energies are measured from the 4(1) state in the reactant complex. The 4(2), 4(3), and 4(4) states are close in energy and lie approximately 60 kcal/mol above the 4(1) state. These high-lying states are unlikely to contribute to the spin inversion through the SOC effect; therefore, we considered only the 4(1)-6 SOC matrix elements. The 4(1)-6 SOC values remain nearly zero during this process; thus, the spin inversion from the 4(1) state to the sextet state hardly occurs in this region.

We have described SOC analyses of the direct conversion of CH_4 to CH_3OH by FeO^+. The reaction pathway involves three main crossing seams between the quartet and sextet potential-energy surfaces. The first crossing seam is located near TS1, which is a transition state for the cleavage of a C–H bond of CH_4. This spin inversion is the most important aspect in this reaction pathway because, in this crossing region, the spin inversion occurs from the sextet state to the quartet state, leading to a substantial decrease in the barrier height of TS1 from 31.1 to 22.1 kcal/mol at the B3LYP level.

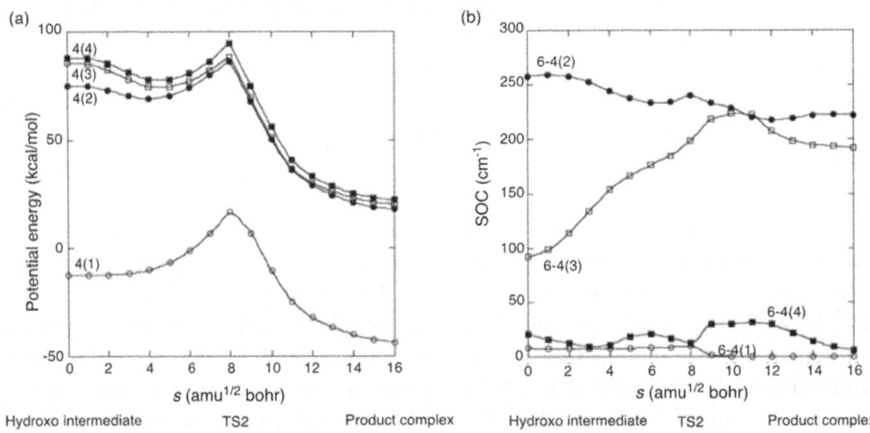

Fig. 9 **a** Potential energies at the CASSCF level and **b** SOC values along the IRC. The IRC connects the reaction pathway from the hydroxo intermediate ($s = 0$) to the product complex ($s = 16$) via TS2 ($s = 8.0$). The 4(1), 4(2), 4(3), and 4(4) states are the first-, second-, third-, and fourth-excited quartet states, respectively. Reproduced from Ref. [37]. Copyright 2003 American Institute of Physics

The second crossing seam occurs in the vicinity of the hydroxo intermediate HO–Fe^+–CH_3. The third crossing seam is located in the exit channel in which CH_3OH is eliminated from the product complex. We calculated the strength of the SOC along the reaction pathway to estimate the probability of spin inversion. The strength of SOC is 133.6 cm^{-1} in the reactant complex and 21.4 cm^{-1} in the hydroxo intermediate, and the SOC value decreases along the reaction pathway, approaching zero in the product complex. Although the strength of SOC decreases as we follow the reaction coordinate, the SOC value in the second crossing seam is not negligible. However, this spin inversion is not preferred for the production of CH_3OH because TS2 in the sextet state is relatively high. When the quartet state changes to the sextet state in the vicinity of the second crossing seam, another reaction branch that leads to $FeOH^+$ + $^\bullet CH_3$ will occur. Finally, we conclude that the spin inversion from the sextet state to the quartet state should occur in the first crossing seam, whereas the spin state should remain unchanged in both the second and third crossing seams. Our computations suggest that the spin inversion can promote the reaction of FeO^+ with CH_4 and that the spin inversion would not occur in the reverse reaction (Fe^+ + CH_4) that starts in the third crossing seam.

5 Summary

In this chapter, we reviewed mechanisms for the direct conversion of CH_4 to CH_3OH by bare transition-metal monoxide cations (MO^+), as deduced from DFT calculations. The calculated electronic structures of the MO^+ ions are important for the reactivity and catalytic properties. DFT calculations indicated that MO^+ (M = Sc, Ti, V, Cr, Mn, Fe, Co, Ni, Cu) reacted with CH_4. Both the high- and low-spin state potential-energy surfaces have been characterized at the B3LYP/6-311G** level of theory. In conclusion, the CH_4–CH_3OH conversion by MO^+ proceeds in the following way.

$$MO^+ + CH_4 \rightarrow OM^+(CH_4) \rightarrow TS1 \rightarrow HO{-}M^+{-}CH_3 \rightarrow TS2$$
$$\rightarrow M^+(CH_3OH) \rightarrow M^+ + CH_3OH$$
$$(M = \text{Sc, Ti, V, Cr, Mn, Fe, Co, Ni, Cu})$$

Potential energies measured from the dissociation limit indicate a substantial activation barrier, consistent with experimental results. The behavior of the forward and reverse reactions in the MO^+ complexes with CH_4 systems is consistent with reactions that proceed via the hydroxo intermediates CH_3–M^+–OH. Ab initio calculations showed that the formation of CH_3OH via CH_3–M^+–OH is energetically more favorable than the byproduct pathway leading to methyl radical, except in the case of ScO^+ and TiO^+.

The potential-energy surface crossing points between the high- and low-spin states play a substantial role, leading to a decrease in the barrier height of transition states.

In the early transition-metal oxides, the crossing point exists prior to TS2, the three-centered transition state for methyl migration. The crossing point of the NiO^+ exists prior to TS1 (the four-centered transition state for the cleavage of a C–H bond of CH_4), and that of CuO^+ exist after TS1. For the NiO^+ potential-energy surface, there is another crossing point in exit channel; this crossing point does not play a substantial role.

Reasonably detailed potential-energy surfaces for this conversion can explain the experimental observations that the early transition-metal oxides hardly react with CH_4, whereas the late transition-metal oxides react with CH_4. In conclusion, the early transition-metal oxides cannot proceed to the product complex via TS2, which is very high in energy, as the rate-determining step. After H-atom abstraction, the late transition-metal oxides proceed to the product complex.

In the FeO^+/CH_4 system, crossing seams along the quartet and sextet IRC paths and the SOC effects for the conversion of CH_4 were calculated. As a result of the IRC path, there are three crossing seams between the sextet and quartet potential-energy surfaces, and spin inversion should occur twice: FeO^+ (sextet) + CH_4 → $[FeO^+(CH_4)]$ (sextet) → spin inversion → TS1 (quartet) → $[HO–Fe^+–CH_3]$ (quartet) → TS2 (quartet) → $[Fe^+(CH_3OH)]$ (quartet) → spin inversion → Fe^+ (sextet) + CH_3OH. The first crossing seam is the most important aspect of this reaction pathway because a spin inversion from the sextet to the quartet state decreases the activation barriers for the C–H bond dissociation of CH_4. From CASSCF wavefunctions, SOC was computed along the reaction pathway to estimate the probability of spin inversion. The SOC value decreases along the reaction pathway, approaching zero in the product complex. The SOC analysis indicated that the spin inversion from the sextet state to the quartet state should occur in the first crossing seam, whereas the spin state should remain unchanged in both the second and third crossing seams.

The reaction in the condensed phase must be considered to predict the activity of catalysts in experiments. The two-step concerted mechanism presented here is expected to not be limited to the hydroxylation of gas-phase CH_4 by MO^+, and these results can be used to understand the catalytic and enzymatic processes concerning alkane hydroxylation. We believe that the results of the gas-phase experiments and calculations provide valuable insight into the elementary step of the oxidation process and that the combination of experimental and theoretical studies enables us to realize the design of effective catalysts for the hydroxylation of alkanes.

Acknowledgements This project was partially supported by JST-CREST "Innovative Catalysts" JPMJCR15P5.

References and Notes

1. The Activation of Saturated Hydrocarbons by Transitions Metal Complexes, Shilov AE (1984) Riedel Publishing: Dordrecht, The Netherlands
2. Shilov AE, Shul'pin GB (1997) Chem Rev 97:2879–2932

3. Shilov AE, Shteinman AA (1999) Acc Chem Res 32: 763–771
4. Activation and Functionalization of Alkanes, Ed: Hill CL (1989) Wiley, New York
5. Selective Hydrocarbon Activation, Davies JA, Watson PL, Liebman JF, Greenberg A (1990) VCH, New York
6. The Organometallic Chemistry of the Transition Metals, Crabtree RH (1990) Wiley, New York
7. Crabtree RH (1985) Chem Rev 85:245–269
8. Crabtree RH (1995) Chem Rev 95:987–1007
9. Arndtsen BA, Bergman RG, Mobley TA, Peterson TH (1995) Acc Chem Res 28:154–162
10. Hall C, Perutz RN (1996) Chem. Rev 96:3125–6146
11. Lunsford JH (1995) Angew Chem Int Ed Engl 34:970–980
12. Armentrout PB (1991) Science 251:175–179
13. Eller K, Schwarz H (1991) Chem Rev 91:1121–1177
14. Roth LM, Freiser BS (1991) Mass Spectrom Rev 10:303–328
15. Weisshaar JC (1993) Acc Chem Res 26:213–219
16. Gesser HD, Hunter NR, Prakash CB (1985) Chem Rev 85:235–244
17. Jackson TC, Jacobson DB, Freiser BS (1984) J Am Chem Soc 106:1252–1257
18. Kang H, Beauchamp JL (1986) J Am Chem Soc 108:5663–5668
19. Kang H, Beauchamp JL (1986) J Am Chem Soc 108:7502–7509
20. Jackson TC, Carlin TJ, Freiser BS (1986) J Am Chem Soc 108:1120–1126
21. Schröder D, Schwarz H (1990) Angew Chem Int Ed Engl 29:1433–1434
22. Schröder D, Fiedler A, Hrusák J, Schwarz H (1992) J Am Chem Soc 114:1215–1222
23. Ryan MF, Fiedler A, Schröder D, Schwarz H (1994) Organometallics 13:4072–4081
24. Ryan MF, Fiedler A, Schröder D, Schwarz H (1995) J Am Chem Soc 117:2033–20140
25. Schwarz J, Wesendrup R, Schröder D, Schwarz H (1996) Chem Ber 129:1463–1474
26. Clemmer DE, Aristov N, Armentrout PB (1993) J Phys Chem 97:544–552
27. Chen Y-M Clemmer DE, Armentrout PB (1994) J Am Chem Soc 116:7815–7826
28. A review on the work of Schwarz' group: Schröder D, Schwarz H (1995) Angew Chem Int Ed Engl 34:1973–1995
29. Azzaro M, Breton S, Decouzon M, Geribaldi S (1993) Int J Mass Spectrom Ion Processes 128:1–20
30. Dietl N, van der Linde C, Schlangen M, Beyer MK, Schwarz H (2011) Angew Chem Int Ed Engl 50:4966–4969
31. Yoshizawa K, Shiota Y, Yamabe T (1997) Chem Eur J 3:1160–1169
32. Yoshizawa K, Shiota Y, Yamabe T (1998) J Am Chem Soc 120:564–572
33. Yoshizawa K, Shiota Y, Yamabe T (1998) Organometallics 17:2825–2831
34. Yoshizawa K, Shiota Y, Yamabe T (1999) J Chem Phys 111:538–545
35. Yoshizawa K, Kagawa, Y, Shiota Y, Yamabe T (2000) J Phys Chem A 104:2552–2561
36. Shiota Y, Yoshizawa K (2000) J Am Chem Soc 122:12317–12326
37. Shiota Y, Yoshizawa K (2003) J Chem Phys 118:5872–5879
38. Yoshizawa K J (1998) Biol Inorg Chem 3:318–324
39. Yoshizawa K (2000) J Inorg Biochem 78:23–34
40. Yoshizawa K, Ohta T, Yamabe T (1998) Bull Chem Soc Jpn 71:1899–1909
41. Yoshizawa K, Yumura T (2003) Chem Eur J 9:2347–2358
42. Huang S-P, Shiota Y, Yoshizawa K (2013) Dalton Trans 42:1011–1023
43. Yoshizawa K, Suzuki A, Shiota Y, Yamabe T (2000) Bull Chem Soc Jpn 73:815–827
44. Yoshizawa K (2002) Coord Chem Rev 226:251–259. https://doi.org/10.1016/S0010-8545(01)00464-7
45. Yoshizawa K, Shiota Y (2006) J Am Chem Soc 128:9873–9881
46. Shiota Y, Yoshizawa K (2009) Inorg Chem 48:838–845
47. Shiota Y, Juhász G, Yoshizawa K (2013) Inorg Chem 52:7907–7917
48. Itoyama S, Doitomi K, Kamachi T, Shiota Y, Yoshizawa K (2016) Inorg Chem 55:2771–2775
49. Miyanishi M, Abe T, Hori Y, Shiota Y, Yoshizawa K (2019) Inorg Chem 58:12280–12288
50. Mahyuddin MH, Staykov A, Shiota Y, Yoshizawa K (2016) ACS Catal 6:8321–8331
51. Mahyuddin MH, Shiota Y, Staykov A, Yoshizawa K (2017) Inorg Chem 56:10370–10380

52. Mahyuddin MH, Staykov A, Shiota Y, Miyanishi M, Yoshizawa K (2017) ACS Catal 7:3741–3751
53. Mahyuddin MH, Shiota Y, Staykov A, Yoshizawa K (2018) Acc Chem Res 51:2382–2390
54. Mahyuddin MH, Tanaka T, Shiota Y, Staykov A, Yoshizawa K (2018) ACS Catal 8:1500–1509
55. Mahyuddin MH, Shiota Y, Yoshizawa K (2018) Catal Sci Tech 8:5875–5885
56. Mahyuddin MH, Shiota Y, Yoshizawa K (2019) Catal Sci Tech 9:1744–1768
57. Carter EA, Goddard WA III (1988) J Phys Chem 92:2109–2115
58. Fiedler A, Schröder D, Shaik S, Schwarz H (1994) J Am Chem Soc 116:10734–10741
59. Clemmer DE, Elkind JL, Aristov N, Armentrout PB (1991) J Phys Chem 95:3387–3393
60. Fisher ER, Elkind JL, Clemmer DE, Georgiadis R, Loh SK, Aristov N, Sunderlin LS, Armentrout PB (1990) 93:2676–2691
61. Lower SK, El-Sayed MA (1966) Chem Rev 66:199–241
62. Bera JK, Samuelson AG, Chandrasekhar J (1998) Organometallics 17:4136–4145
63. Danovich D, Shaik S. (1997) J Am Chem Soc 119:1773–1786
64. Filatov M, Shaik S (1998) J Phys Chem A 102:3835–3846
65. Schröder D, Shaik S, Schwarz H (2000) Acc Chem Res 33:139–145

Enzymatic Methane Hydroxylation: sMMO and pMMO

Takashi Yumura, Takehiro Ohta, and Kazunari Yoshizawa

Abstract Catalytic function of methane monooxygenase (MMO) for the conversion of methane to methanol at ambient pressure and temperature has fascinated many researchers. This chapter will review quantum chemical approaches, especially by means of density functional theory (DFT) calculations, to elucidate mechanisms for the methane hydroxylation by active sites of MMO (soluble and particulate MMOs) with a focus on how a methane C–H bond is activated. Because of the uncertainty in the coordination environment of transition metal cations in the active site that has a direct reactivity toward substrate methane, different computational models in various spin states have been used. Reflecting from these differences, a variety of mechanisms for the methane hydroxylation on MMO has been proposed: radical-related mechanisms (radical-rebound and nonsynchronous concerted mechanisms) and nonradical mechanism. These proposed mechanisms will be compared. In addition, key factors in determining the reactivity of MMO will be discussed on the basis of molecular orbital theory applied to transition-metal containing active sites.

Keywords Density functional theory calculations · Activation of a methane C–H bond · Spin states · Reaction mechanism · Coordination environment

---

T. Yumura (✉)
Faculty of Materials Science and Engineering, Graduate School of Science and Technology, Kyoto Institute of Technology, Matsugasaki, Sakyo-ku, Kyoto 606-8585, Japan
e-mail: yumura@kit.ac.jp

T. Ohta
Department of Applied Chemistry, Faculty of Engineering, Sanyo-Onoda City University, Yamaguchi 756-0884, Japan

K. Yoshizawa
Institute for Materials Chemistry and Engineering, Kyushu University, 744 Motooka, Nishi-ku, Fukuoka 819-0395, Japan

K. Yoshizawa (ed.), *Direct Hydroxylation of Methane*,
https://doi.org/10.1007/978-981-15-6986-9_3

1 Experimental Background

Methane monooxygenase (MMO) [1–9], found in methanotroph bacteria, catalyzes the transformation of methane and dioxygen into methanol and water at ambient pressure and temperature. Since methane is an inert hydrocarbon due to high dissociation energy of its C–H bond (104 kcal/mol), the catalytic function of MMO has fascinated many researchers, and led them to investigate how methane is converted into methanol under physiological condition. MMO has two different forms: cytoplasmic soluble MMO (sMMO) [1–9] and membrane bound particulate MMO (pMMO) [10]. To clarify mechanisms for the methane to methanol conversion by MMO, it is important to obtain structural information of active sites responsible for the methane hydroxylation. The first X-ray crystal structures of active sites in sMMO and pMMO were reported in 1993 [9] and 2005 [10], respectively. The sMMO form consists of three protein components [11–14]: a dimeric $\alpha_2\beta_2\gamma_2$ hydroxylase (MMOH), and NADH-dependent [Fe_2S_2] and FAD-containing reductase (MMOR), and regulatory protein (MMOB). A carboxylate-bridged diiron center is contained in each a subunit of MMOH as the active site for the activation of dioxygen, and methane hydroxylation [9, 15–19]. On the other hand, the pMMO form is a trimer with an $\alpha_3\beta_3\gamma_3$ polypeptide arrangement [10]. In soluble regions of each α (pmoB) subunit, mononuclear and dinuclear coppers are contained, and zinc is located within the membrane as third metal center. The X-ray crystallographic analyses suggested that iron and copper centers play an important role in the direct methane hydroxylation by sMMO or pMMO, respectively [9, 10].

The catalytic cycle for sMMO has been well investigated. In a proposed catalytic cycle (Fig. 1), the resting state of MMOH is the oxidized form ($MMOH_{ox}$), that contains two Fe(III) cations in an octahedral coordination environment [9, 18] (Fig. 2a). The diiron(III) state converts to the reduced Fe(II)Fe(II) state ($MMOH_{red}$) by accepting electrons provided from a reductase protein MMOR [20, 21]. During the reduction process, a carboxylate shift of Glu243 occurs at the active site of $MMOH_{red}$, where five-coordinate iron cations appear [18, 19] (Fig. 2b). Because of the presence of a vacant coordination site of the diiron center of $MMOH_{red}$, this intermediate interacts with dioxygen to form peroxo intermediates ($MMOH_{peroxo}$). As shown in Fig. 1, the peroxo intermediates, which were spectroscopically monitored [22–27], subsequently converts into a bright-yellow intermediate with the high-valent Fe(IV)Fe(IV) state [28, 29]. This intermediate, labeled Q ($MMOH_Q$) has a direct reactivity toward substrate methane. Combined Mössbauer and EXAFS investigations suggested that the active site of $MMOH_Q$ should involve a $(\mu\text{-O})_2Fe(IV)_2$ diamond core (Fig. 1), whose iron cations are antiferromagnetically coupled [30]. In addition, EXAFS analyses indicated that the coordination number of the iron cations is no greater than 5 [30].

It is important to elucidate how the $(\mu\text{-O})_2Fe(IV)_2$ active site of $MMOH_Q$ oxidizes methane into methanol. One possible mechanism for the methane hydroxylation by MMOH is radical rebound mechanism [31], proposed by an analogy to the same reaction by cytochrome P-450, whose active site involves a mononuclear iron-oxo

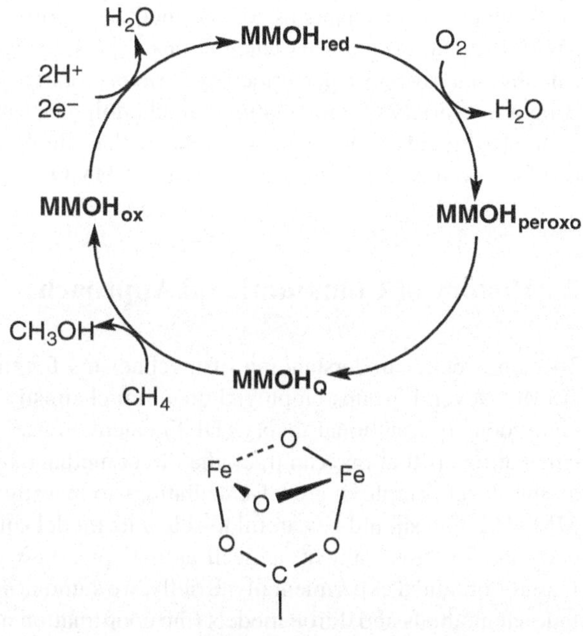

Fig. 1 A proposed change in the catalytic cycle of methane hydroxylation by sMMO. A proposed structure of an intermediate contributing to direct hydroxylation of methane into methanol is also displayed. Reproduced from Ref. [52] by permission of Wiley Ltd.

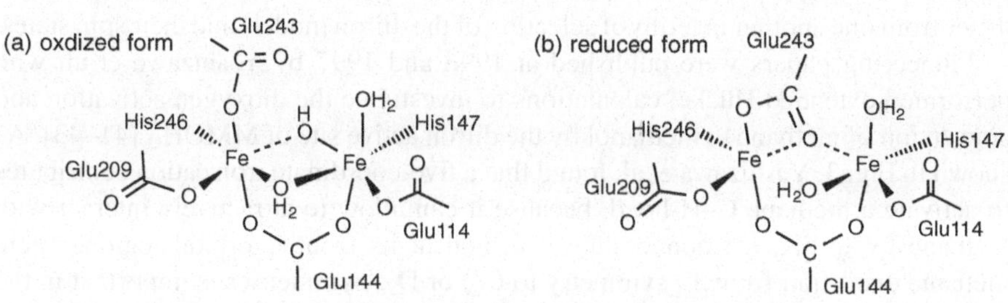

Fig. 2 X-ray crystal structures of sMMO in the oxidized (**a**) and reduced (**b**) forms of diiron

species coordinated by a porphyrin ring [32, 33]. In this mechanism, the homolytic dissociation of a methane C–H bond results in the formation of a methyl radical, and an OH group in the active site of $MMOH_Q$. After that, the methyl radical is rebound to the formed OH group to form a methanol complex. However, this proposal has been disputed by the following experimental reports. As one experimental report, Newcomb and Lippard used radical-clock substrates to be reacted with MMOH [34–39]. As a result, they found a quite shorter lifetime of a putative radical species formed in the hydroxylation reaction (≈150 fs), which is inconsistent with the formation of a discrete radical species. A similar short lifetime for the radical species was also observed in the hydroxylation of chiral ethane on MMOH [40]. Despite many experimental investigations, the mechanism for the C–H bond activation of methane by $MMOH_Q$ has been under debate.

To elucidate mechanisms for the methane hydroxylation by sMMO as well as pMMO, quantum chemistry calculations, especially density functional theory (DFT) calculations have been performed by various research groups. In this chapter, we will compare a variety of computational methodologies using different models of MMO applied to elucidate mechanisms for the methane hydroxylation with an aim to clarify key factors in determining the reactivity of MMO.

2 History of Computational Approach

To gain a better understanding of mechanisms for the methane hydroxylation by sMMO, several groups employed quantum chemistry calculations [41–101], especially density functional theory (DFT) calculations. Advantage of the use of DFT calculations is that one can track the direct methane hydroxylation by $MMOH_Q$ on atomic-level. Employing DFT calculations to investigate the methane activation by $MMOH_Q$, one should be careful to select its model suitable to describe the iron(IV) coordination environment, as well as the spin state suitable to describe magnetic features obtained experimentally. Briefly, we summarize in Table 1 details of computational methods and diiron models (the coordination number of the dual iron cation, and the spin state) in published papers. Table 1 shows that these computational studies differ from one another in terms of selection of the diiron models and their spin states.

Pioneering papers were published at 1996 and 1997 by Yoshizawa et al. who performed extended Hückel calculations to investigate the dioxygen activation and conversion of methane to methanol by the diiron active site of $MMOH_Q$ [41–43]. As shown in Fig. 3, Yoshizawa et al. found that a five-coordinate iron cation contributes to activate a methane C–H bond, because it can allow to attractively interact with methane by utilizing a nonbonding d orbital at its frontier orbital region. Then, methane distortion from T_d symmetry to C_{3v} or D_{2d} symmetries is important in the attractive orbital interactions [43]. This orbital interaction results in the formation of a direct Fe–C coupling, which facilitates the electron transfer from the highest occupied molecular orbital (HOMO) of methane (C–H bonding character) to the unoccupied nonbonding d orbital [46, 102]. This electron transfer induces to activate a methane C–H bond. The findings obtained from extended Hückel calculations are qualitative [41–43], however, they can provide fundamental insights on roles of the orbital interactions in the activation of a methane C–H bond.

The publication by Yoshizawa et al. stimulated to quantitatively study the methane to methanol conversion by $MMOH_Q$ by means of density functional theory (DFT) calculations. Representative models used in the papers are displayed in Fig. 4. At 1997, DFT calculations with B3LYP functional (DFT/Hartree–Fock hybrid functional) [103–107] were carried out by Yoshizawa et al. [44, 46] and Siegbahn et al. [57] who used diiron models, whose iron centers are coordinated by formate, water, and OH groups. See models labeled as (a), (b), and (k) in Fig. 4. Unfortunately, these models are too small to represent the coordination sphere of iron cations of sMMO. These studies were followed by Morokuma group at 1999 [66] and Friesner group

Table 1 Key factors in catalytic activity of models of active sites of $MMOH_Q$ for the cleavage of a methane C–H bond used in literatures

References	Method	Cluster model[a]	Spin states	CN of Fe[b]	Active site[c]	C–H bond cleavage	
						Energy barrier[d]	Fashion[e]
Yoshizawa group							
[46]	DFT (B3LYP)	(a, b)	^{1}A	4, 5	Fe	4.3	Nonradical
[52]	DFT (B3LYP)	(c)	^{1}A (IS-AF)	5, 6	Fe	30.4	Nonradical
[56]	DFT (B3LYP*)	(d)	HS-AF/HS-FM[f]	6, 6	B–O	25.2	Radical
		(e)	HS-AF/HS-FM[f]	6, 6	B–O	23.5	
		(f)	HS-AF/HS-FM[f]	4, 6	B–O	17.4[g]	
		(g)	HS-AF/HS-FM[f]	5, 6	B–O	19.4[g]	
		(h)	HS-AF/HS-FM[f]	6, 6	B–O	22.6[g]	
		(i)	HS-AF/HS-FM[f]	6, 6	T–O	21.9[g]	
		(j)	HS-AF/HS-FM[f]	4, 6	B–O	14.3[g]	
Siegbahn group							
[57]	DFT (B3LYP)	(k)	^{9}A	4, 5	B–O	4.5	Radical
[59]	DFT (B3LYP)	(l)	^{9}A	6, 6	B–O	7.1	Radical
[63]		(m)	^{9}A	6, 6	B–O	13.8	Radical
Morokuma group							
[66]	DFT (B3LYP)	(n)	$^{9}A/^{11}A$[h]	6, 6	B–O	19.5	Radical
[68]	DFT (B3LYP)	(n)	$^{9}A/^{11}A$[h]	6, 6	B–O	23.2 (O-side) 19.5 (N-side)	Radical
[74]	DFT (B3LYP)	(o)	$^{9}A/^{11}A$[h]	6, 6	B–O	21.8	Radical

(continued)

Table 1 (continued)

References	Method	Cluster model[a]	Spin states	CN of Fe[b]	Active site[c]	C–H bond cleavage	
						Energy barrier[d]	Fashion[e]
Friesner group							
[77, 83]	DFT (B3LYP)	(p)	^{1}A (HS-AF)	6, 6	B–O	17.9	Radical
[87]	QM/MM (B3LYP)	–	^{1}A (HS-AF)	6, 6	B–O	17.9	Radical

[a]Cluster model labels are displayed in Fig. 4.
[b]CN of Fe: Coordination number of iron cations in models of the active site of $MMOH_Q$
[c]Active site: an atom that directly activates a methane C–H bond, iron cation (Fe), bridging oxygen atom (B–O) or terminal oxygen atom (T–O)
[d]Energy barrier: energy barrier for dissociation of a methane C–H bond in kcal/mol
[e]Fashion for dissociation of a methane C–H bond: how a methane C–H bond is activated via non-radical or radical fashions. Detailed definition can be seen in Fig. 5
[f]In Ref. [56], two spin states were considered: HS-AF or HS-FM states. In the initial state, the ground state is the HS-AF state
[g]Spin crossing occurs before the step for the C–H bond activation. The barrier was estimated based on the transition state in the HS-FM state
[h]In Ref. [68], two spin states were considered: ^{9}A or ^{11}A states. In the initial state, the ground state is the ^{9}A state. In this study spin crossing was not found before the step for the C–H bond activation
[i]In Ref. [68], a bridging oxygen atom in H_2O or NH_2 sides (denoted by O-side or N-side) activates a methane C–H bond

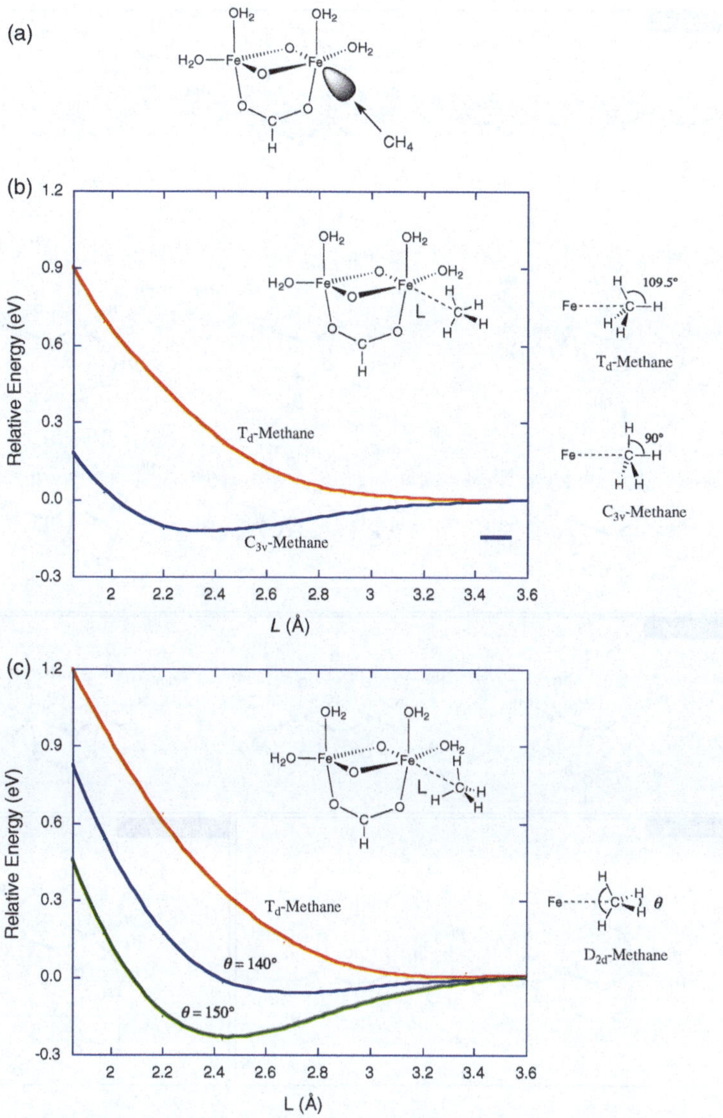

Fig. 3 **a** Interaction between CH_4 and coordinatively unsaturated diiron active site of $MMOH_Q$ with a nonbonding orbital in its frontier orbital region. **b** total energy diagrams (from extend Hückel calculations) for the coordination of T_d- and C_{3v}-type methanes to a coordinatively unsaturated diiron active site of $MMOH_Q$ as a function of the separation between the iron and methane carbon atoms (L). Similar energy diagram in the D_{2d}-type methane case can be seen in (**c**). Reprinted with the permission from Ref. [43]. Copyright 1997 American Chemical Society

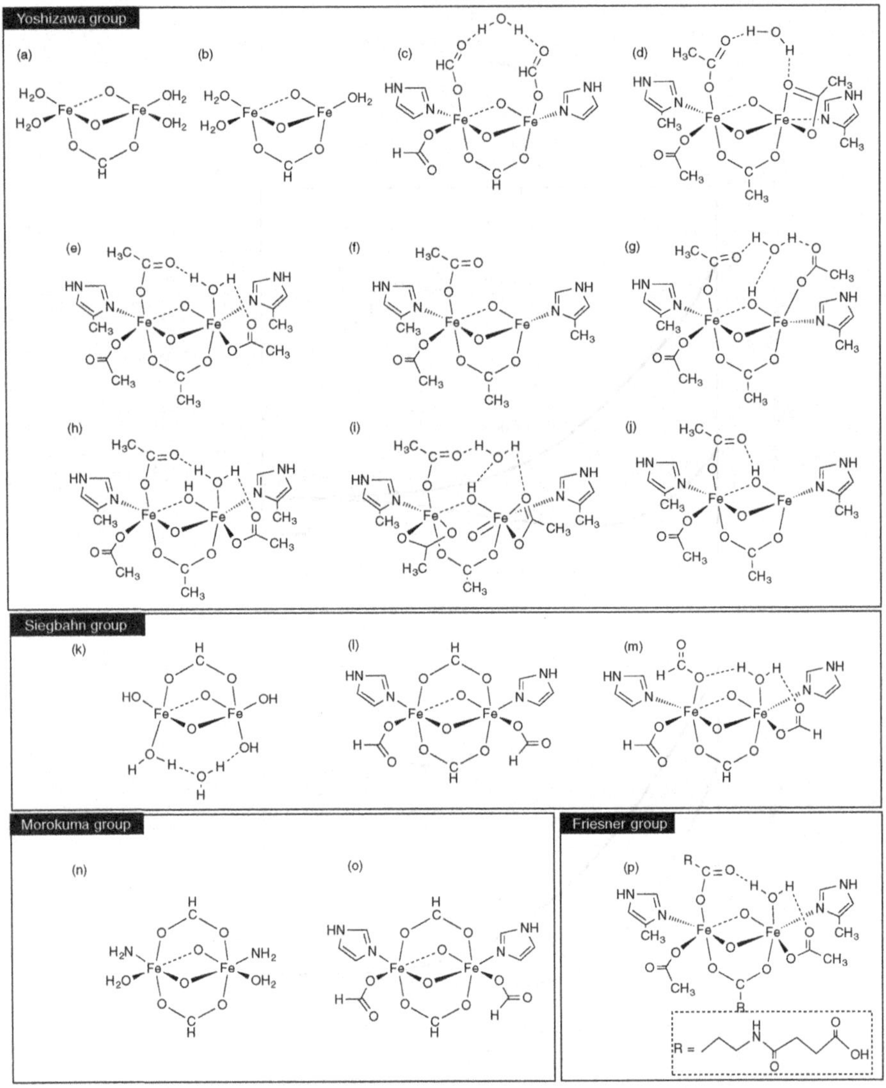

Fig. 4 Various diiron models of $MMOH_Q$ used in literatures. Models of (**a**) and (**b**) were taken from Ref. [46], that of (**c**) from Ref. [52], those from (**d**) to (**j**) from Ref. [56], that of (**k**) from Ref. [57], that of (l) from Ref. [59], that of (**m**) from Ref. [63], that of (n) from Refs. [66] and [68], that of (**o**) from Ref. [74], and that of (**p**) from Refs. [77] and [83]

at 2000 [76, 77] who used models labeled as (n) and (p) in Fig. 4, respectively. The diiron models used after 2000 (Fig. 4c–j, l, m, o, p) contain ligands more realistic to mimic actual glutamate and histidine residues [52, 53, 56, 59, 63, 74]. The improvement, where the first coordination sphere around the diiron active site was reasonably modeled, increased the reliability of data obtained from DFT calculations, such as bonding and the energy barrier for the activation of a methane C–H bond, to compare them with corresponding experimental values.

A further improvement in studying the reactivity of $MMOH_Q$ toward methane is to incorporate entire protein of sMMO. In this direction, Noodlemann et al. investigated at 2001 the structure of active site in sMMO by considering electrostatic interactions between the active site and the protein environment, and found that the protein plays an influential role in determining the structure and the energetics of the active site [90–93]. After the development in the computation of the enzymatic reactions, Morokuma et al. [73] directly incorporated protein environments by using mixed quantum mechanics/molecular mechanics (QM/MM) ONIOM approaches [108, 109]. ONIOM calculations found that the protein environments control local structures of the active sites of $MMOH_{ox}$ and $MMOH_{red}$. After that, Friesner group also performed QM/MM calculations containing ~7200 atoms to investigate the potential energy surface of the methane to methanol conversion by $MMOH_Q$ [87, 88].

3 Key Factors in Determining Reactivity of $MMOH_Q$ Toward Methane

First, let us discuss key factors in determining the reactivity of $MMOH_Q$, summarized in Table 1. One is the coordination number of iron cations in the active site of $MMOH_Q$, and the other is its spin state. Although EXAFS studies [30] suggested that $MMOH_Q$ has an iron cation whose coordination number is no greater than 5, its exact structure has not been clear. Because of the uncertainty in the coordination number of iron cations in $MMOH_Q$, different computational models have been proposed. Models that contain octahedrally-coordinated diiron species were used by many research groups (e.g. Siegbahn [59, 63], Morokuma [66, 68, 74], and Friesner [77, 83] groups). In contrast, Yoshizawa group used a different model containing 4- or 5-coordinated diiron species [44, 46, 52, 53, 56], putting considerable emphasis on the EXAFS findings.

As is well known, the coordination number of iron cations determines their *d*-orbital splitting. When an iron atom is octahedrally coordinated, the relevant *d* orbitals are splitting into three low-lying t_{2g} orbitals and two highly-lying e_g orbitals [110]. If $MMOH_Q$ involves a kind of edge-shared six-coordinate diiron(IV) complex where iron–iron interaction is weak, energy levels of their d-based frontier orbitals are basically described in Fig. 5a. Since formal charge of each iron cation in $MMOH_Q$ is +4, each iron cation has four *d*-electrons. Accordingly $MMOH_Q$ has eight *d*-electrons that are housed on the split *d* orbitals. Depending on how eight electrons are housed on the split d orbitals, there are at least four types of spin states in $MMOH_Q$, as depicted in Fig. 5b–e. In ^9A or ^5A, four parallel spins (high-spin (HS)) or two parallel spins (intermediate-spin (IS)) on both iron cations are ferromagnetically coupled, respectively. On the other hand, antiferromagnetically coupled diiron(IV)

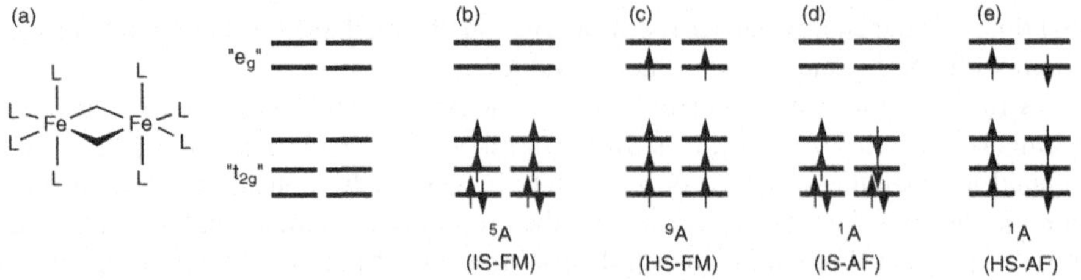

Fig. 5 The electronic states of coordinatively saturated diiron active site of $MMOH_Q$ with eight *d* electrons; *d*-orbital splitting and four possible spin sates. Iron cations in an active site have high-spin (HS) or intermediate-spin (IS) state. An active site of $MMOH_Q$ with ferromagnetically coupled iron cations has ^5A or ^9A states described by IS-FM or HS-FM, respectively. As active site of $MMOH_Q$ with antiferromagnetically coupled irons has open-shell singlet states described by IS-AF or HS-AF

active sites results in the two types of open-shell single state; one has high-spin on each iron cation, and the other has intermediate-spin on each iron cation.

Mössbauer analysis indicated that $MMOH_Q$, being EPR inactive, contains antiferromagnetically coupled diiron(IV) species [30], and therefore calculations of $MMOH_Q$ in open-shell singlet spin states are necessary. For this purpose, Siegbahn [59, 63], Friesner [77, 83], Noodleman [94, 96, 97], Yamaguchi [98], and Yoshizawa [52, 56] groups used broken-symmetry (BS) approach [111–113] to describe the observed antiferromagnetic states of the active site in $MMOH_Q$. The BS approach is based on unrestricted DFT calculations for open-shell singlet states where spin-up and spin-down electrons are allowed to be localized on different atomic centers. Open-shell spin states obtained from the BS approach are not a pure spin state described by a single determinant, but rather than a weighted average of pure spin states to give a spin-coupling pattern with such an antiferromagnetic alignment. However, this approach is accepted as a good representation of the non-orthogonal, natural magnetic orbitals in dinuclear complexes. In contrast, Morokuma et al. [66, 68, 74] and Siegbahn et al. [59, 63] used ^9A and ^{11}A spin states for $MMOH_Q$ to study the methane hydroxylation, instead of the open-shell singlet, because they assumed that antiferromagnetic and ferromagnetic couplings do not affect the reaction mechanism due to small exchange couplings.

Here let us closely look at models of the active site of $MMOH_Q$ in Fig. 4. Figure 6 displays spin density distributions on representative diiron models. The $MMOH_Q$ models in Fig. 4 can be categorized into three groups in Chart 1: two models contain a $(\mu\text{-}O)_2Fe(IV)_2$ core, and the other contains a $(\mu\text{-}O)(\mu\text{-}OH)Fe(IV)Fe(III)$ core [56]. The two $(\mu\text{-}O)_2Fe(IV)_2$ models can be distinguished by whether a coordinatively unsaturated iron cation exists or not. Coordinatively saturated $(\mu\text{-}O)_2Fe(IV)_2$ models were used by Siegbahn [59, 63], Morokuma [66, 68, 74], and Friesner [77, 83] groups. Siegbahn and Morokuma groups adopted the nonet and undecet spin states for the diiron models labeled by (l and m) and (n and o) in Fig. 4, respectively. According to their calculations, the bridging oxygen atoms in the diamond core have significant spin densities in the high-spin sates, as seen in Fig. 6d–g. In contrast,

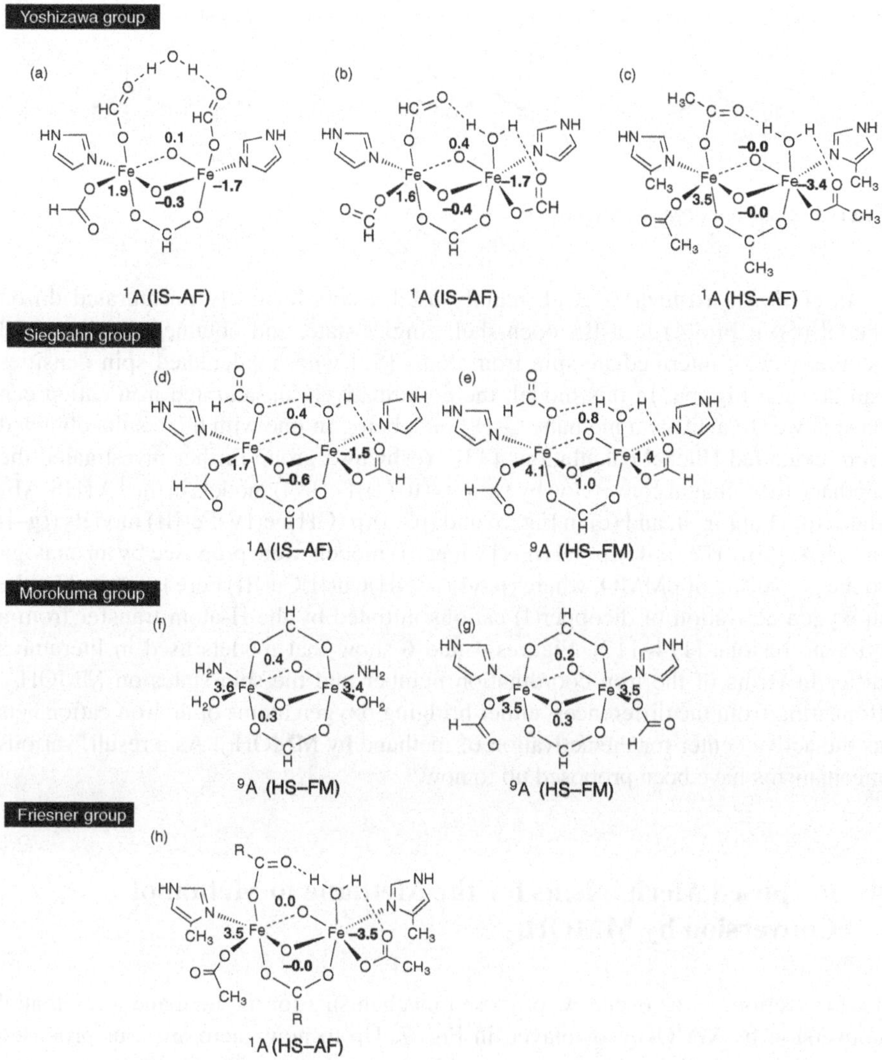

Fig. 6 Calculated atomic spin densities for representative diiron active site of $MMOH_Q$ in different spin states. Data in (**a**) and (**b**) were taken from Ref. [52], that in (**c**) from Ref. [56], that in (**d**) and (**e**) from Ref. [63], that in (**f**) from Ref. [66], that in (**g**) from Ref. [74], and that in (**h**) from Refs. [77] and [83]

Friesner et al. adopt a BS open-shell singlet spin state for the coordinatively saturated $(\mu\text{-O})_2Fe(IV)_2$ model labeled by (p) in Fig. 4. Their calculations found negligible spin densities on bridging oxygen atoms in the diamond core in the open-shell singlet state, as seen in Fig. 6h. In the coordinatively saturated $(\mu\text{-O})_2Fe(IV)_2$ models, the iron cations cannot bind into methane for the activation of its C–H bond. Instead, a bridging oxygen atom can serve as an active center for the C–H bond activation.

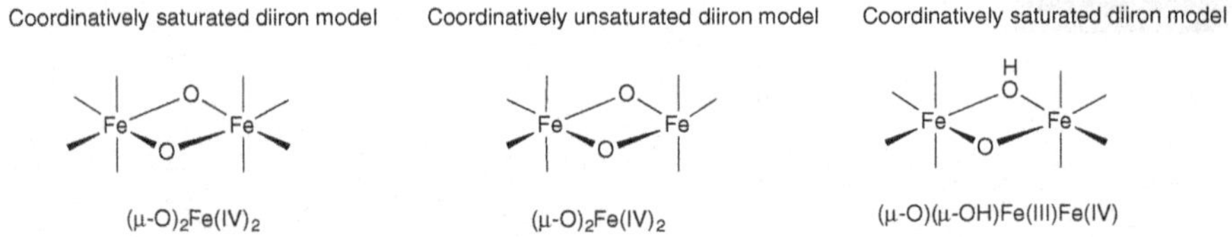

Chart 1 Catagory of $MMOH_Q$ models

In contrast, Yoshizawa et al. mainly used a coordinatively unsaturated diiron model (c) in Fig. 4) in a BS open-shell singlet state, and obtained its optimized geometry with intermediate-spin iron atoms [52] whose calculated spin densities can be seen Fig. 6a. In this model, the coordinatively unsaturated iron cation can have power to activate a methane C–H bond, being in line with the results obtained from extended Hückel calculations [43]. Yoshizawa group further investigated the methane to methanol conversion by various $(\mu\text{-O})_2Fe(IV)_2$ models in the ^1A (HS-AF) state ((d–f) in Fig. 4, and (c) in Fig. 6) and (μ-O)(μ-OH)Fe(IV)Fe(III) models ((g–j) in Fig. 4) [56]. The (μ-O)(μ-OH)Fe(IV)Fe(III) models were proposed by an analogy to the active site of pMMO, where (μ-O)(μ-OH)Cu(II)Cu(III) core is formed by the dioxygen activation of dicopper(I) cations initiated by the H-atom transfer from a tyrosine residue [114–117]. Figures 4 and 6 show that models used in literatures differ in terms of the iron coordination number and the spin states on $MMOH_Q$. Reflecting from the differences, either bridging oxygen atoms or an iron cation acts as the active center for the activation of methane by $MMOH_Q$. As a result, various mechanisms have been proposed up to now.

4 Proposed Mechanisms for the Methane to Methanol Conversion by $MMOH_Q$

In this section, let us overview proposed mechanisms for the methane to methanol conversion by $MMOH_Q$, displayed in Fig. 7. Up to now, there are four proposed mechanisms, which can be distinguished by how a methane C–H bond is cleaved in the reaction course. In the three mechanisms in Fig. 7a–c, the activation of a methane C–H bond by a bridging oxygen atom of the coordinatively saturated $(\mu\text{-O})_2Fe(IV)_2$ active site proceeds in a homolytic manner. After the homolytic dissociation of a methane C–H bond, there are three types of reaction pathways. One mechanism proceeds in line with the radical rebound mechanism, whose detailed steps were explained above (Fig. 7a) [59, 63, 66, 68, 74]. This radical rebound mechanism was theoretically investigated by Siegbahn and Morokuma groups. In another mechanism proposed by Crabtree and Siegbahn, displayed in Fig. 7b, the methyl radical recombines with a coordinatively unsaturated iron atom via a weak Fe–CH_3 bond after the H-atom abstraction [57, 58]. The remaining mechanism starting from the

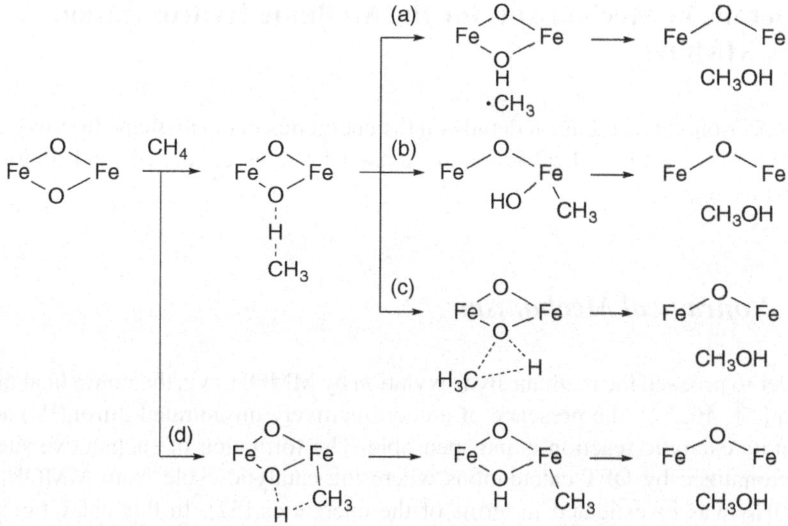

Fig. 7 Four possible pathways for methane hydroxylation by $MMOH_Q$. Reproduced from Ref. [52] by permission of Wiley Ltd.

homolytic dissociation of a methane C–H bond is a nonsynchronous concerted mechanism (Fig. 7c) proposed by Friesner group [77, 83]. After the homolytic cleavage of a methane C–H bond, the resultant methyl moiety does not detach from the hydoxo group, and then methanol is formed in a nearly barrierless fashion.

In contract to the radical-related mechanisms, Yoshizawa group proposed a nonradical mechanism for the methane hydroxylation by $MMOH_Q$, whose active site was assumed to contain a coordinatively unsaturated iron cation [44, 46, 52], as shown in Fig. 7d. In the first step of the mechanism, the coordinatively unstaturated iron atom can activate a C–H methane bond by utilizing a direct Fe–C coupling. The C–H bond activation of methane results in the formation of a HO–Fe–CH_3 species. After that, the binding of the resultant methyl ligand and the hydoxo ligand to form a methanol complex. A striking feature of this mechanism is that neither radical species nor ionic species are involved in the course of the hydroxylation reaction, because the methyl moiety formed by the C–H bond dissociation is directly bound to the coordinatively unsaturated iron atom. The nonradical mechanism was proposed by an analogy to the methane hydroxylation by bare FeO^+ species, which is generated by the reaction of Fe^+ and pulsed-in N_2O under Fourier transform ion cyclotron resonance (FTICR) conditions [118–121].

5 Details in Mechanisms for the Methane Hydroxylation by $MMOH_Q$

In this section, let us focus on details on the energetics in the methane hydroxylation by $MMOH_Q$ via the radical rebound, nonsynchronous concerted, and nonradical mechanisms.

5.1 Nonradical Mechanism

In order to proceed the methane hydroxylation by $MMOH_Q$ via the nonradical mechanism [44, 46, 52], the presence of a coordinaitively unsaturated diiron(IV) active site in its catalytic reaction is indispensable. The formation of such active site can be rationalized by DFT calculations where the catalytic cycle from $MMOH_{red}$ to $MMOH_Q$ was investigated in terms of the energetics [52]. In this catalytic cycle, dioxygen is incorporated into the diiron active site of $MMOH_{red}$, which results in the formation of peroxo intermediates ($MMOH_{peroxo}$) and $MMOH_Q$. According to the observed X-ray structure of $MMOH_{red}$, Ref. [52] assumed that its active site consists of five-coordinate iron cations, before the reaction with dioxygen, as shown in Fig. 8. In the optimized model of $MMOH_{red}$, a formate ligand forms a bidentate chelating interaction with the left-hand iron atom and a single bond with the right-hand iron atom. Each iron atom of $MMOH_{red}$ has a five-coordinate environment, and thus the active site directly interacts with dioxygen to form peroxo intermediates with different dioxygen binding modes ($MMOH_{peroxo}$) After that, the peroxo intermediates can convert into $MMOH_Q$ with a five-coordinate iron atom. In fact, the coordinatively unsaturated diiron active site of $MMOH_Q$ is 56.3 kcal/mol stable relative to the dissociation limit toward $MMOH_{red}$ and dioxygen. This result strongly suggested the formation of a coordinatively unsaturated diiron(IV) active site of $MMOH_Q$ in the catalytic cycle is feasible from a viewpoint of the energetics [52].

After the formation of the coordinatively unsaturated diiron active site, the five-coordinate iron center can utilize its nonbonding d orbital to directly interact with methane to form a methane complex In the methane complex, attractive interactions can operate between methane and the coordinatively unsaturated diiron active site, which is in contrast to the other radical-related mechanisms, as will be described. These interactions through direct Fe–C couplings facilitate to abstract an H atom from methane via a four-centered transition state (TS1). This process requires an activation energy of 30.4 kcal/mol relative to the dissociation limit toward methane and $MMOH_Q$ [52]. The intramolecular H atom migration leads to the formation of an important intermediate where newly formed OH and CH_3 groups directly coordinate to an iron atom as ligands. This intermediate, called hydoxo intermediate, corresponds to the important insertion species ($HO–Fe^+–CH_3$) formed in the methane to methanol conversion by FeO^+ in the gas phase [118–121].

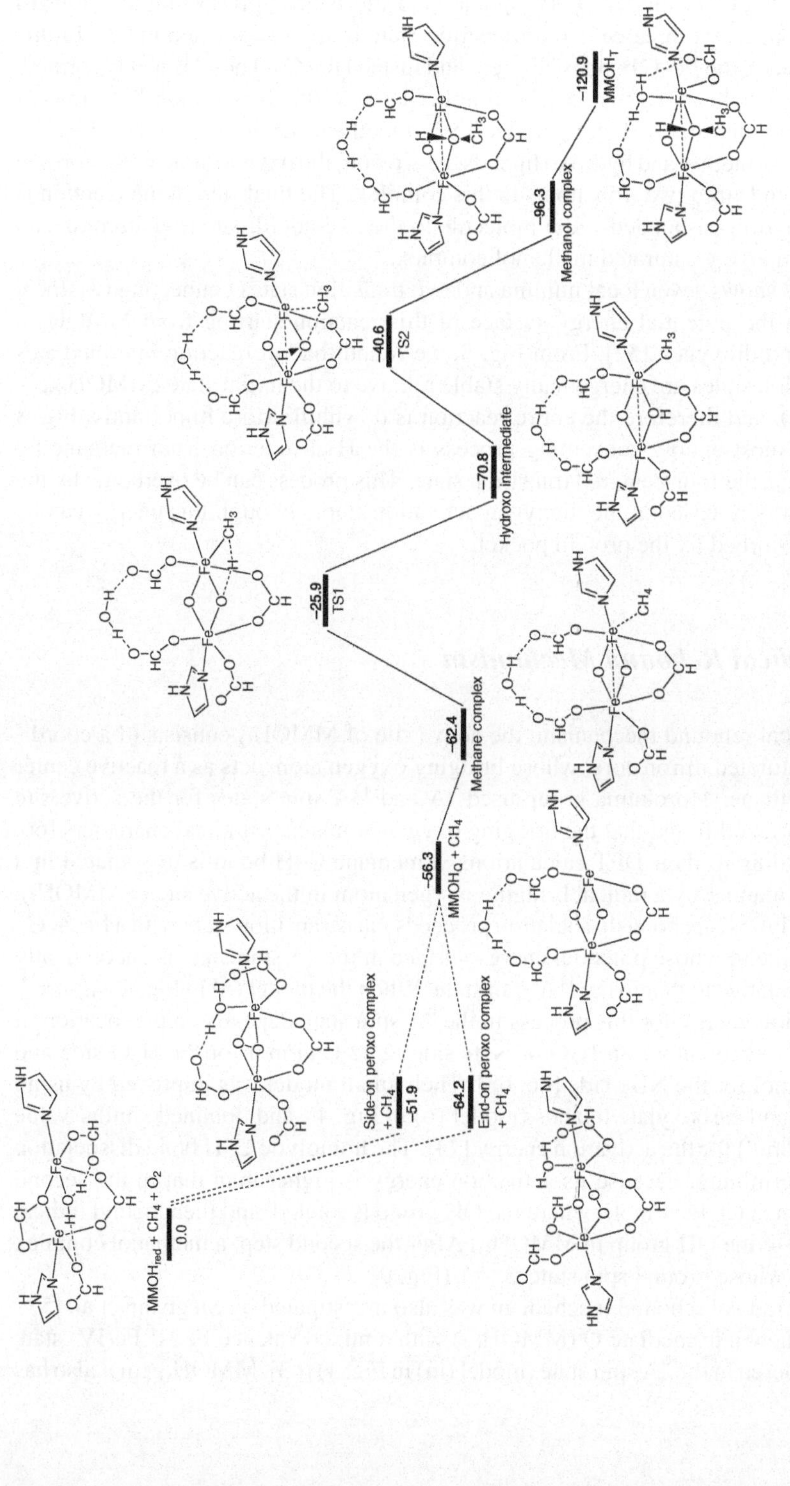

Fig. 8 Potential energy diagram for dioxygen activation and methane hydroxylation by $MMOH_Q$ in the broken-symmetry singlet state at B3LYP calculations. Energy relative to $MMOH_{red} + O_2 + CH_4$ are in kcal/mol. $MMOH_Q$ has coordinatively unsaturated diiron active site corresponding to that in Fig. 4c. Reproduced from Ref. [52] by permission of Wiley Ltd.

After the formation of the hydoxo intermediate, the hydrogen atom of the OH group rotates out of the $(\mu\text{-O})_2Fe_2$ plane, and the methyl group migrates toward the OH group via a three-centered transition state in the second step (TS2). In this transition state, the Fe–C bond is cleaved, and instead the C–O bond is newly formed. The barrier height of TS2 is 15.7 kcal/mol relative to the dissociation limit toward methane and the intermediate Q. After TS2, a methanol complex is formed by the binding of the methyl and hydroxo ligands. As a result, the oxygen atom of the formed methanol binds into two iron atoms in this complex. The final step in the reaction is returning hydrogen-bonded water molecule to the five-coordinate iron atom to lead to a coordinatively saturated methanol complex.

Figure 8 shows seven local minima and two transition states connecting two local minima on the potential energy surface of this reaction starting from $MMOH_{red}$, methane and dioxygen [52]. From Fig. 8, we found that all reaction intermediates and transition states are energetically stable relative to the initial state ($MMOH_{red}$ + $CH_4 + O_2$), and therefore the entire reaction is downhill. More importantly, Fig. 8 shows the most energy-consuming process is the H-abstraction from methane by $MMOH_Q$ via the four-centered transition state. This process can be overcome by the excess energy released in the dioxygen activation step, although this energy can be partially absorbed by the protein pocket.

5.2 *Radical Rebound Mechanism*

In the radical rebound mechanism, the active site of $MMOH_Q$ consists of a coordinatively saturated diiron core, whose bridging oxygen atom acts as a reactive center toward methane. Morokuma group used ^9A and ^{11}A spin states for the active site of $MMOH_Q$, and found that the bridging oxygen atoms have radical characters [66, 68]. According to their DFT calculations, a methane C–H bond is dissociated in a homolytic manner by a radical bridging oxygen atom in the active site of $MMOH_Q$. This homolytic C–H bond dissociation proceeds via a transition state with a linear C–H–O alignment, whose potential energy surface in the ^9A spin state is energetically favorable relative to that in the ^{11}A spin state. When the model (*n*) in Fig. 4 was used, the activation energy for this process in the ^9A spin state depends on the location of a bridging oxygen atom on H_2O or NH_2 side: 23.2 kcal/mol for the H_2O-side and 19.5 kcal/mol for the NH_2-side [66, 68]. Their small model was improved by using imidazole and carboxylate ligands (model (o) in Fig. 4), and obtained similar value (21.8 kcal/mol) for the activation energy [74]. The homolytic C–H bond dissociation is rate-determining, because its activation energy is higher than that in the second step (less than ten kcal/mol), where the OH group is rotated, and then methyl radical is rebound to the OH group in $MMOH_Q$. After the second step, a methanol complex is formed, whose ground spin state is ^{11}A (Fig. 9).

Similar radical rebound mechanism was also investigated by Siegbahn et al. [57–65] But a new intermediate Q ($MMOH_{Q'}$) with a mixed valence Fe(III)Fe(IV) state was considered in the ^9A spin state (model (m) in Fig. 4) [63]. $MMOH_{Q'}$, that also has

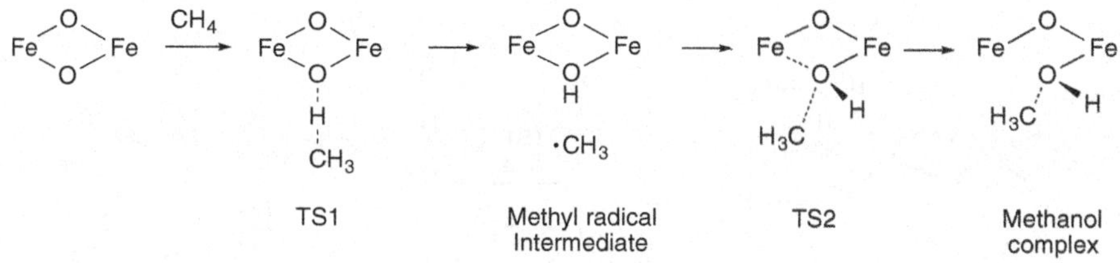

Fig. 9 Detailed reaction pathway of the methane to methanol conversion by coordinatively saturated diiron active site of $MMOH_Q$ via a radical rebound mechanism

a radical bridging oxygen atom, is a few kcal/mol unstable relative to $MMOH_Q$ with the Fe(IV)Fe(IV) state. The radical bridging oxygen atom can dissociate a methane C–H bond in a homolytic manner, whose activation energy (13.8 kcal/mol) is smaller than those obtained by Morokuma group. Although the different value was obtained for the activation energy for the homolytic C–H bond dissociation, the mechanism proposed by Siegbahn et al. was essentially the same as that by Morokuma et al. It should be noted that there is a drawback in using the active sites in the ^9A and ^{11}A states, because such ferromagnetic states cannot be observed from Mössbauer and EPR experiments [30].

To overcome the drawback, Friesner [77] and Yoshizawa groups [56] used active sites of $MMOH_Q$ with an antiferromagnetically coupled diiron(IV) cations to investigate whether it can activate a methane C–H bond in a homolytic manner. Their studies obtained the active sites in open-shell singlet states by the BS approach, and found negligible spin densities on their bridging oxygen atoms than those in the ferromagnetic spin states. The bridging oxygen atom can also activate a methane C–H bond in a homolytic manner by utilizing a structural change of the Fe_2O_2 core upon the methane binding at the initial step of the reaction course. The activation energies for the homolytic C–H bond dissociation by $MMOH_Q$ in the open-shell states were calculated to be 17.9 and 23.5 kcal/mol by Frisener et al. and Yoshizawa et al. who used Jaguar and Gaussian programs, respectively. See Fig. 10. Note that the structural distortion triggers an intramolecular electron transfer from a bridging oxygen atom to an iron cation, forming a radical on the bridging atom [83]. However, this structural distortion of $MMOH_Q$ destabilizes a methane complex, being different from that in nonradical mechanism. Furthermore, Yoshizawa group investigated energy profiles of the methane to methanol conversion by $(\mu\text{-O})_2Fe(IV)_2$ and (μ-O)(μ-OH)Fe(III)Fe(IV) active sites with different iron coordination environment (models (d–j) in Fig. 4) [56]. As a result of DFT calculations, the (μ-O)(μ-OH)Fe(III)Fe(IV) active sites are more reactive toward methane than the $(\mu\text{-O})_2Fe(IV)_2$ active sites, as shown in Table 1, where the activation energies for the methane C–H bond dissociation by $MMOH_Q$ models are tabulated. In particular, a four-coordinated (μ-O)(μ-OH)Fe(III)Fe(IV) model is the most effective for the methane hydroxylation, as shown in Fig. 11, where energetically equivalent radical and non-radical intermediates exist.

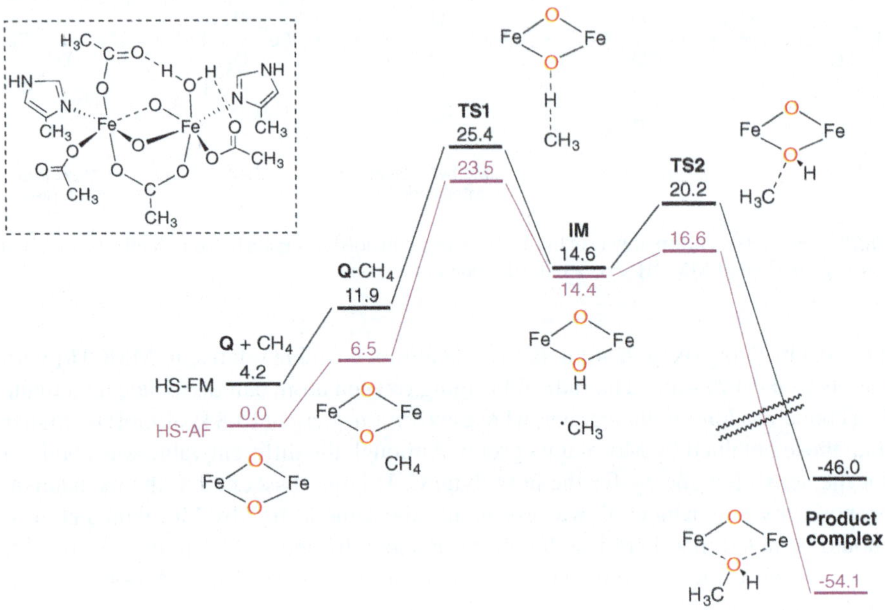

Fig. 10 Calculated free energy profiles of the reactions of methane with the coordinatively saturated diiron active site of $MMOH_Q$ in the HS-FM and HS-AF states at the B3LYP level, where a methane C–H bond is activated via a homolytic manner. The active site model corresponds to that in Fig. 4e. The relative free energy in kcal/mol. Reproduced from Ref. [56] with permission from The Royal Society of Chemistry

5.3 Nonsynchronous Concerted Mechanism

Friesner and Lippard proposed that two possible reaction branches open after the homolytic dissociation of a methane C–H bond by $MMOH_Q$, whose active site model has an open-shell singlet state [76–89]. The model is given by (*p*) in Fig. 4. One reaction branch follows the radical rebound mechanism, involving an intermediate where a methyl radical separates by 1.97 Å from the OH ligand. In their computations, a slightly smaller activation barrier (3.9 kcal/mol) was obtained than that in Ref. [74]. The other reaction branch involves a slight different intermediate where a methyl radical does not detach from the OH group (Fig. 12). In fact, the separation between the methyl radical and OH ligand is 1.54 Å. After the formation of the new intermediate, methanol is formed in a nearly barrierless fashion. This new reaction branch is called as nonsynchronous concerted mechanism.

Because of a negligible activation energy for the methanol formation step (1.3 kcal/mol) [77], Friesner et al. suggested that the nonsynchronous concerted mechanism can explain a shorter lifetime of radicals formed in the relevant experiments [34–39]. In contrast, a longer radical lifetime is expected in the radical rebound mechanism, whose methanol formation step requires a higher activation energy.

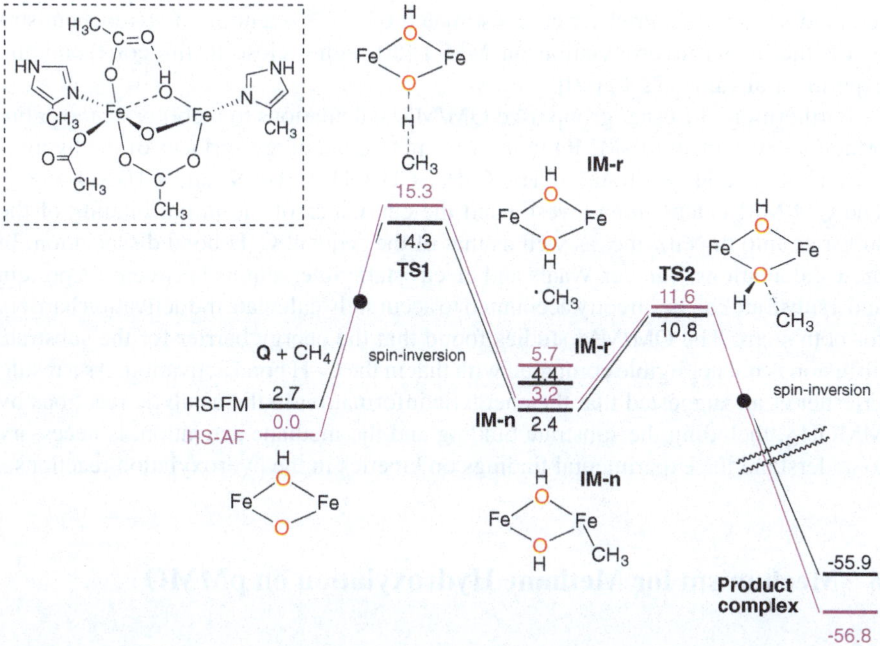

Fig. 11 Calculated free energy profiles of the reactions of methane with the coordinatively unsaturated diiron active site of $MMOH_Q$ in the HS-FM and HS-AF states at the B3LYP level, where a methane C–H bond is activated via a homolytic manner. The active site model corresponds to that in Fig. 4j. The relative free energy in kcal/mol. Reproduced from Ref. [56] with permission from The Royal Society of Chemistry

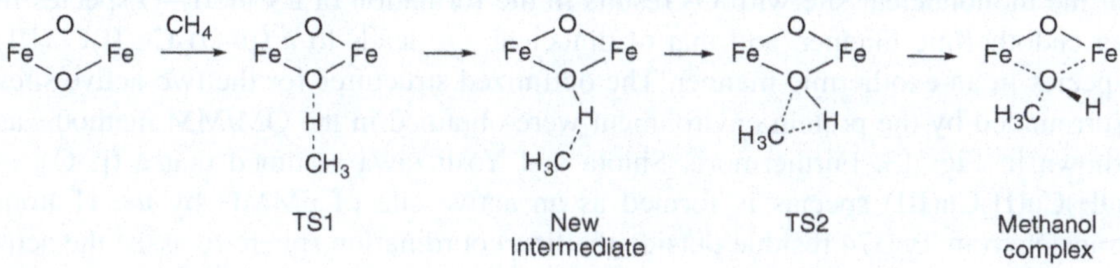

Fig. 12 Detailed reaction pathway of the methane to methanol conversion by coordinatively saturated diiron active site of $MMOH_Q$ via a non-synchronous concerted mechanism

Stereochemistry for the hydroxylation of a chiral ethane on MMOH was further investigated from a semi classical molecular dynamic (MD) simulation [80]. According to their semiclassical MD studies [80], the nonsynchronous concerted mechanism yields the product alcohol for chiral ethane with the total retention of configuration. On the other hand, a longer lifetime of the ethyl radical in the radical rebound mechanism can allow to rotate its C–C bond, yielding the product alcohol with an inversion of configuration. Considering a mixture of nonsynchronous concerted and radical rebound trajectories, Friesner et al. estimated 69–84% retention of stereochemistry of for the ethane hydroxylation on MMO [80] being close to the corresponding experimental value (72%) [40].

Furthermore, Friesner group used QM/MM calculations to incorporate the entire protein environment of sMMO to improve potential energy surfaces of the hydroxylation of several substrates (CH_4, C_2H_6, CH_3OH, CH_3CN, and CH_3NO_2) [87]. The QM/MM calculations investigated the energetics of the initial binding of the substrate into the enzyme, as well as that of the sequent C–H bond dissociation. In these calculations, van der Waals and electrostatic interactions between the protein and a substrate can be directly accounted to accurately calculate the activation barriers for both steps. The QM/MM studies found that the energy barrier for the substrate diffusion is not negligible compared with that in the C–H bond activation. As a result, Friesner et al. suggested that the energetic information in the catalytic reactions by $MMOH_Q$, including the substrate binding and the methane activation, is necessary to understand the experimental findings on kinetics in the hydroxylation reactions.

6 Mechanism for Methane Hydroxylation on pMMO

After obtaining the X-ray crystal structure of the particulate form of MMO (pMMO) in 2.8 Å resolution [10], Shiota and Yoshizawa performed DFT and QM/MM calculations to elucidate its catalytic function for the methane hydroxylation [114–117]. Taking the crystal structure of pMMO into account, mononuclear and dinuclear copper species, located within the soluble regions of the pmoB subunit, were considered as the active site for the methane to methanol conversion [114]. A reaction of the mononuclear site with O_2 results in the formation of a Cu(III)–O species in an endothermic manner, and that of dinuclear site leads to a $(\mu\text{-O})_2$Cu(II)Cu(III) species in an exothermic manner. The optimized structures for the two active sites surrounded by the protein environment were obtained in the QM/MM methods, as shown in Fig. 13. Furthermore, Shiota and Yoshizawa assumed that a (μ-O)(μ-OH)Cu(II)Cu(III) species is formed as an active site of pMMO by the H-atom transfer from Tyr374 residue outside the first coordination sphere to assist the activation of the O–O bond [116]. The formation of the (μ-O)(μ-OH)Cu(II)Cu(III) active site was proposed by an analogy to the oxidation of tyrosine to dopaquinone by tyrosinase [122]. On the other hand, Chen et al. proposed a $(\mu_3\text{-O})$trinuclear Cu(II)Cu(II)Cu(III) active site for the methane hydroxylation, which proceeds via a concerted oxo-insertion manner [123, 124]. However, such a trinuclear copper site was not observed in the X-ray crystal structure of pMMO [10].

Here, let us focus on mechanisms for the methane hydroxylation by the Cu(III)–O, $(\mu\text{-O})_2$Cu(II)Cu(III), and (μ-O)(μ-OH)Cu(II)Cu(III) active sites of pMMO. By looking at the coordination environment of a copper cation in the QM/MM optimized active sites [114] as well as the X-ray crystal structure of pMMO [10], model complexes were set for DFT calculations to search the methane to methanol conversion. The first model is the Cu(III)–O species coordinated by two imidazole and

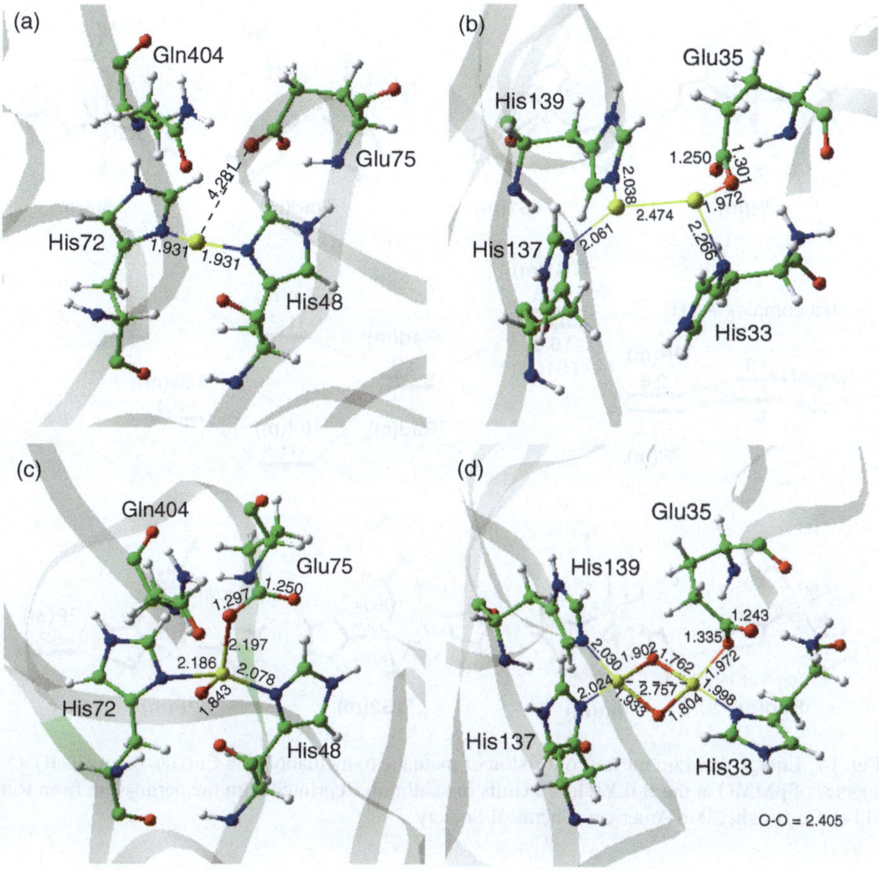

Fig. 13 QM/MM-optimized structures for **a** the monocopper site, **b** the dicopper site, **c** the Cu(III)–O (or Cu(II)–O$^{\cdot}$) species, and **d** the (μ-O)$_2$Cu(II)Cu(III) species of pMMO. Reprinted with the permission from Ref. [114]. Copyright 2006 American Chemical Society

one acetate, and the second model is the (μ-O)$_2$Cu(II)Cu(III) species coordinated by three imidazole and one acetate [114]. The third model consists of the (μ-O)(μ-OH)Cu(II)Cu(III) species, where one copper cation is coordinated by imidazole, and the other cation is coordinated by the N-terminal to the main chain of a histidine residue [116]. By using the active site models of pMMO, potential energy surfaces of the methane hydroxylation were obtained at B3LYP level of theory.

Potential energy surfaces of the methane hydroxylation by the Cu(III)–O, (μ-O)$_2$Cu(II)Cu(III), and (μ-O)(μ-OH)Cu(II)Cu(III) active sites are displayed in Figs. 14, 15, and 16, respectively. Similar reaction mechanisms were proposed, independent of the three types of active site [114, 116]. At the first state of the reaction, methane is weakly bound to a copper cation in an active site to form a methane complex. After that, an oxo group can dissociate a methane C–H bond in a transition

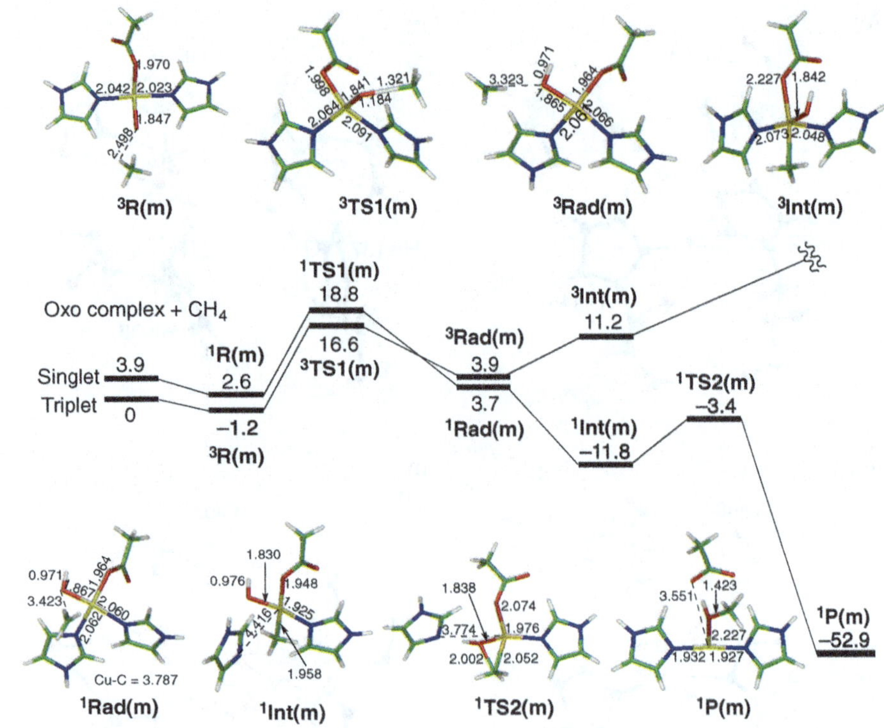

Fig. 14 Energy diagram for the conversion of methane to methanol by a Cu(III)–O (or Cu(II)–O˙) species of pMMO at the B3LYP level. Units in kcal/mol. Reprinted with the permission from Ref. [114]. Copyright 2006 American Chemical Society

state with a linear C–H–O alignment. The resultant methyl radical is directly bound to a copper cation to form a nonradical intermediate. After the formation of the nonradical intermediate, the second half of this reaction is the recombination between the OH and CH_3 ligands via a triangle-shape transition state. The recombination process yields a methanol complex.

Despite the same reactions mechanisms, different potential energy surfaces of the methane hydroxylation were obtained by using the three active site models. In the Cu(III)–O active site whose ground state is spin triplet, the potential energy surfaces in the triplet and singlet spin states are energetically close [114] (Fig. 14). In this case, the C–H bond dissociation proceeds on the triplet potential energy surface, whose the activation energy was calculated to be 17.8 kcal/mol relative to the methane complex. After that, binding of methyl radical into the copper cation forms a nonradical intermediate, where the copper cation has a formal charge of +1. The nonradical intermediate is extremely stable in the singlet spin state, because the contained copper cation has a d^{10} system. Starting from the nonradical intermediate, the second half of this reaction proceeds in the single state. The activation energy for the second half of this reaction is lower than that for the first half, and therefore

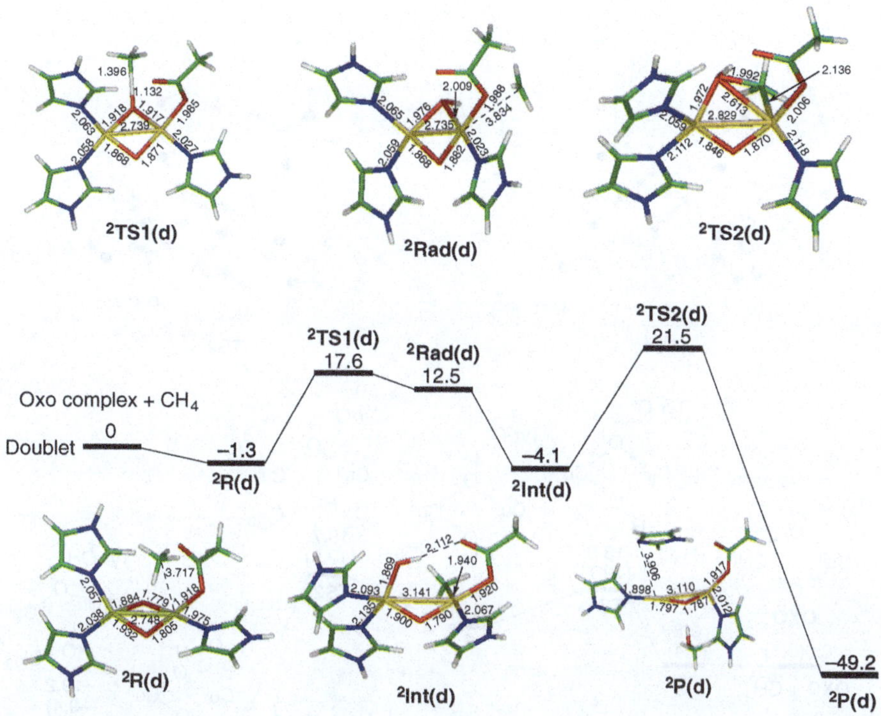

Fig. 15 Energy diagram for the conversion of methane to methanol by the (μ-O)$_2$Cu(II)Cu(III) species of pMMO in the doublet state at the B3LYP level. Units in kcal/mol. Reprinted with the permission from Ref. [114]. Copyright 2006 American Chemical Society

the rate-limiting step in this reaction pathway is in the C–H bond dissociation step. Accordingly, DFT calculations found that two spin states participate in the methane hydroxylation, where spin-crossing from the triplet to singlet state takes place after the C–H bond dissociation step. Note that similar spin-crossing, where spin-orbital coupling plays an important role, can be seen in various types of transition-metal mediated reaction [125–128].

In contrast, the methane hydroxylation by the (μ-O)$_2$Cu(II)Cu(III), and (μ-O)(μ-OH)Cu(II)Cu(III) active sites of pMMO proceeds on surfaces with uniform spin multiplicity, as shown in Figs. 15 and 16, respectively [114, 116]. In the (μ-O)$_2$Cu(II)Cu(III) active site, the methane hydroxylation proceeds on the doublet potential energy surface, because it surface is quite lower in energy than the quartet surface. The activation energy for the C–H bond dissociation was calculated to be 18.9 kcal/mol measured from the methane complex, and that for the recombination step was 25.6 kcal/mol from the nonradical intermediate. Thus, the rate-limiting step is the recombination step, which unfortunately cannot explain large kinetic isotope effects observed experimentally. In contrast, Shiota and Yoshizawa estimated the activation energy for the C–H bond dissociation (16.1 kcal/mol) by the

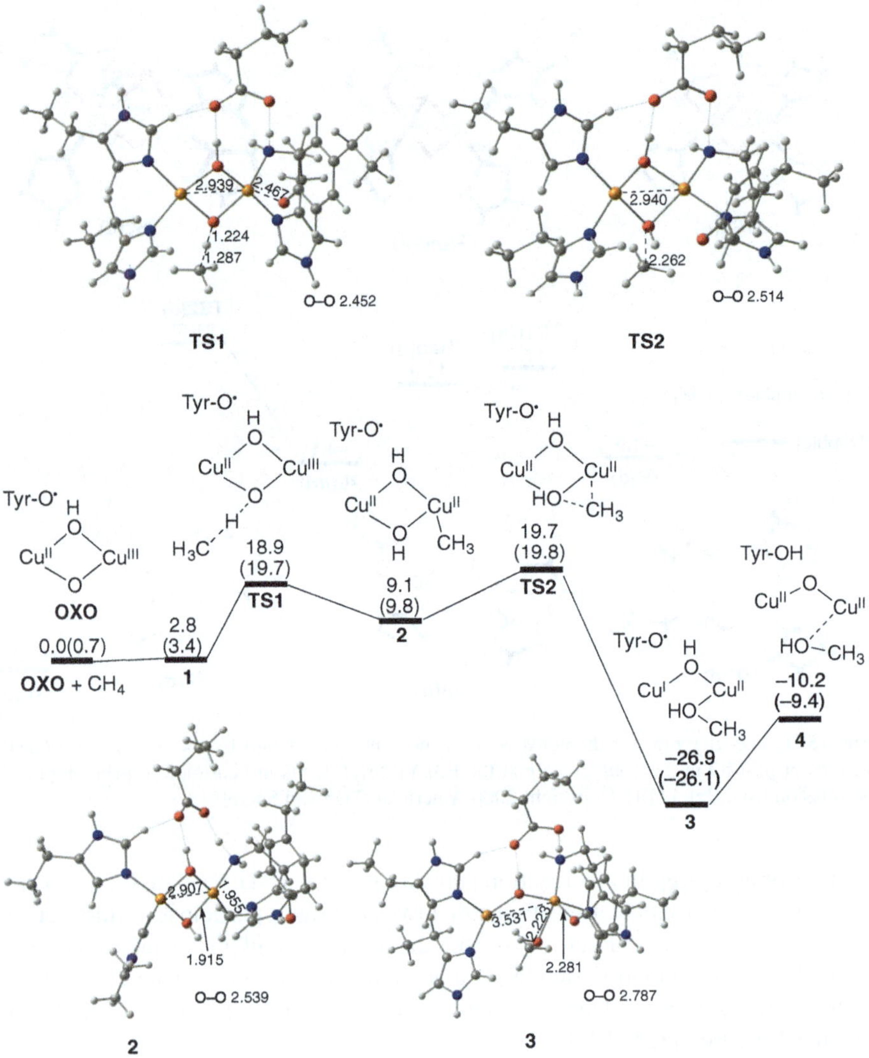

Fig. 16 Energy diagram for the conversion of methane to methanol by the (μ-O) (μ-OH)Cu(II)Cu(III) species of pMMO in the triplet (singlet) state at the B3LYP level. Units in kcal/mol. Reprinted with the permission from Ref. [116]. Copyright 2013 American Chemical Society

(μ-O)(μ-OH)Cu(II)Cu(III) active site, being higher than that in the recombination step (10.6 kcal/mol). DFT calculations suggested that the Tyr374 residue plays an important role in enhancing the reactivity of the bis(μ-oxo) dicopper active site due to an H-atom transfer from the residue. The above DFT findings on the methane hydroxylation on pMMO are also important to give a useful hint to understand reaction mechanisms for the same reaction catalyzed by copper-containing zeolites [129–134].

7 Conclusions, Emerging Issues, and Challenges

This chapter reviewed mechanisms for the direct conversion of methane to methanol on methane monooxygenase (soluble and particulate forms, (sMMO and pMMO)) obtained from density functional theory (DFT) calculations by various research groups. The coordination number of iron cations in the active site of sMMO exhibiting a direct reactivity toward substrate methane ($MMOH_Q$) is uncertain. Accordingly, various groups used different models of the active site in various spin states. As the active site of $MMOH_Q$, coordinatively saturated and unsaturated $(\mu\text{-O})_2Fe(IV)_2$ models where iron cations are ferromagnetically or antiferromagnetically coupled were considered. Reflecting from different coordination environment of an iron cation in the active site models, various mechanisms for the methane hydroxylation by $MMOH_Q$ have been proposed. In coordinatively saturated diiron models, a bridging oxygen atom can activate a methane C–H bond in a homolytic manner to form a methyl radical, together with the hydroxyl group bound to the diiron cations. After the homolytic C–H bond activation, two reaction pathways open. One is a radical rebound mechanism where a methyl radical, depart from the active site, is rebound to the formed hydroxyl group. The other is nonsynchronous concerted mechanism where methyl radical that does not detach from the hydoxo group, and then binding of a methyl group into the hydroxyl group proceeds in a barrierless fashion. In contrast, coordinatively unsaturated diiron models have a vacant coordination site on an iron cation, and it can be utilized to directly interact with methane through the Fe–C bond formation. The direct Fe–C interactions facilitate the activation of a methane C–H bond via a four-centered transition sate to form a $HO\text{–}Fe\text{–}CH_3$ intermediate. After the formation of the intermediate, the methyl group binds into the hydroxyl group to form a methanol complex. Potential energy surfaces along these mechanisms are completely different. Especially, different activation barriers for the C–H bond dissociation were obtained.

Similar mechanistic analyses have be done for the methane hydroxylation on pMMO whose active sites were assumed to be Cu(III)–O, $(\mu\text{–O})_2Cu(II)Cu(III)$, and $(\mu\text{-O})(\mu\text{-OH})Cu(II)Cu(III)$ structures. According to the DFT calculations, the methane hydroxylation on pMMO proceeds via a nonradical intermediate, independent of the three types of active site. A further improvement in studying the reactivity of MMO toward methane is to incorporate its entire protein by means of mixed quantum mechanics/molecular mechanics (QM/MM) calculations. QM/MM calculations found that protein environment has an influential impact on the potential energy surface of this reaction, because the total energies of transition states and intermediates are affected by van der Waals and electrostatic interactions with the protein. The above findings from the sophisticated computational approaches will help to understand the experimental findings on the hydroxylation reactions.

Despite increasing our understanding on the methane hydroxylation on MMO, there still remains issues. Reliability of the computational models relies on the molecular structures developed by spectroscopic and X-ray crystallographic analyses. Recent revisiting studies of the intermediate Q by X-ray absorption spectroscopy (XAS) [135, 136] indicated that the previously proposed structure may be incorrect and revised models have been proposed. However, again the new proposal by XAS is incompatible with the resonance Raman study on the catalytic cycle of sMMO [137]. The structure of the copper active site of pMMO is under debate as well. While the initial X-ray crystal structure showed a dinuclear copper center, recent analyses indicated that a mononuclear copper is involved in the active site [138]. Thus, it remains challenges for computational chemists to perform further analyses on the follow up information.

Spectroscopic and X-ray crystallographic investigations of MMO have provided us valuable information for theoretical studies of ours and others. However still, a question arises as to if a major species that is visible by spectroscopy and X-ray crystallography is really a functional species or not. We cannot eliminate a possibility that a minor speices that escapes from experimental characterizations is a critical component for function of the enzymes. Thus, further efforts should be paid for computational studies to find a truly reactive species responsible for methane oxidation.

Acknowledgements This project was partially supported by a Grant-in-Aid for Scientific Research (C) (No. 18K04864) from JSPS for T.Y.

References and Notes

1. Feig AL, Lippard SJ (1994) Chem Rev 94:759–805
2. Lipscomb JD (1994) Annu Rev Microbiol 48:371–399
3. Liu KE, Lippard SJ (1995) Adv Inorg Chem 42:263–289
4. Wallar BJ, Lipscomb JD (1996) Chem Rev 96:2625–2658
5. Que Jr. L, Dong Y (1996) Acc Chem Res 29:190–196
6. Valentine AM, Lippard SJ (1997) J Chem Soc Dalton Trans 3925–3931
7. Kurtz Jr. DM, (1997) J Biol Inorg Chem 2:159–167
8. Merkx M, Kopp DA, Sazinsky MH, Blazyk JL, Müller J, Lippard SJ (2001) Angew Chem Int Ed 40:2782–2807
9. Rosenzweig AC, Frederick CA, Lippard SJ, Nordlund P (1993) Nature 366:537–543
10. Lieberman RL, Rosenzweig AC (2005) Nature 434:177–182
11. Colby J, Dalton H (1978) Biochem J 171:461–468
12. Woodland MP, Dalton H (1984) J Biol Chem 259:53–60
13. Fox BG, Froland WA, Dege JE, Lipscomb JD (1989) J Biol Chem 264:10023–10033
14. Gassner GT, Lippard SJ (1999) Biochemistry 38:12768–12785
15. Woodland MP, Patil DS, Cammack R, Dalton H (1986) Biochim Biophys Acta 873:237–242
16. Ericson A, Hedman B, Hodgson KO, Green J, Dalton H, Bentsen JG, Beer RH, Lippard SJ (1988) J Am Chem Soc 110:2330–2332
17. Fox BG, Surerus KK, Münck E, Lipscomb JD (1988) J Biol Chem 263: 10553–10556

18. Rosenzweig AC, Nordlund P, Takahara PM, Frederick CA, Lippard SJ (1995) Chem Biol 2:409–418
19. Whittington DA, Lippard SJ (2001) J Am Chem Soc 123:827–838
20. Lund J, Dalton H (1985) Eur J Biochem 147:291–296
21. Lund J, Woodland J, Dalton H (1985) Eur J Biochem 147:297–305
22. Liu KE, Wang D, Huynh BH, Edmondson DE, Salifoglou A, Lippard SJ (1994) J Am Chem Soc 116:7465–7466
23. Liu KE, Valentine AM, Qiu D, Edmondson DE, Appelman EH, Spiro TG, Lippard SJ (1995) J Am Chem Soc 117:4997–4998
24. Liu KE, Valentine AM, Wang D, Huynh BH, Edmondson DE, Salifoglou A, Lippard SJ (1995) J Am Chem Soc 117:10174–10185
25. Lee S-K, Lipscomb JD (1999) Biochemistry 38:4423–4432
26. Valentine AM, Stahl SS, Lippard SJ (1999) J Am Chem Soc 121:3876–3887
27. Brazeau BJ, Lipscomb JD (2000) Biochemistry 39:13503–13515
28. Lee S-K, Nesheim JC, Lipscomb JD (1993) J Biol Chem 268:21569–21577
29. Lee S-K, Fox BG, Froland WA, Lipscomb JD, Münck E (1993) J Am Chem Soc 115:6450–6451
30. Shu L, Nesheim JC, Kauffmann K, Münck E, Lipscomb JD, Que Jr. L (1997) Science 275:515–518.
31. Priestley ND, Floss HG, Froland WA, Lipscomb JD, Williams PG, Morimoto H (1992) J Am Chem Soc 114:7561–7562
32. Cytochrome P450: Structure, Mechanism, and Biochemistry, 2nd ed. (Ed.: Ortiz de Montellano PR (1995) Plenum, New York
33. Sono M, Roach MP, Coulter ED, Dawson JH (1996) Chem Rev 96:2841–2888
34. Liu KE, Johnson CC, Newcomb M, Lippard SJ (1993) J Am Chem Soc 115:939–947
35. Choi S-Y, Eaton PE, Hollenberg PF, Liu KE, Lippard SJ, Newcomb M, Putt DA, Upadhyaya SP, Xiong Y (1996) J Am Chem Soc 118:6547–6555
36. Valentine AM, LeTadic-Biadatti MH, Toy PH, Newcomb M, Lippard SJ (1999) J Biol Chem 274:10771-10776
37. Choi S-Y, Eaton PE, Kopp DA, Lippard SJ, Newcomb M, Shen R (1999) J Am Chem Soc 121:12198-12199
38. Jin Y, Lipscomb JD (1999) Biochemistry 38:6178–6186
39. Brazeau BJ, Austin RN, Tarr C, Groves JT, Lipscomb JD (2001) J Am Chem Soc 123:11831–11837
40. Valentine AM, Wilkinson B, Liu KE, Komar-Panicucci S, Priestley ND, Williams PG, Morimoto H, Floss HG, Lippard SJ (1997) J Am Chem Soc 119:1818–1827
41. Yoshizawa K, Hoffmann R (1996) Inorg Chem 35:2409–2410
42. Yoshizawa K, Yamabe T, Hoffmann R (1997) New J Chem 21:151–161
43. Yoshizawa K, Ohta T, Yamabe T, Hoffmann R (1997) J Am Chem Soc 119:12311–12321
44. Yoshizawa K, Yokomichi Y, Shiota Y, Ohta Y, Yamabe T. (1997) Chem Lett 587–588
45. Yoshizawa K (1998) J Bio Inorg Chem 3:318–324
46. Yoshizawa K, Ohta T, Yamabe T (1998) Bull Chem Soc Jpn 71:1899–1909
47. Yoshizawa K, Suzuki A, Yamabe T (1999) J Am Chem Soc 122:5266–5273
48. Yoshizawa K, Suzuki A, Shiota Y, Yamabe T (2000) Bull Chem Soc Jpn 73:815–827
49. Yoshizawa K (2000) J Inorg Biochem 78:23–34
50. Yoshizawa K (2001) J Organomet Chem 635:100–109
51. Yoshizawa K (2002) Coord Chem Rev 226:251–259
52. Yoshizawa K, Yumura T (2003) Chem Eur J 9:2347–2358
53. Yumura T, Yoshizawa K (2004) Bull Chem Soc Jpn 77:1305–1311
54. Yoshizawa K (2006) Acc Chem Res 39:375–382
55. Yoshizawa K (2013) Bull Chem Soc Jpn 86:1083–1116
56. Huang SP, Shiota Y, Yoshizawa K (2013) Dalton Trans 42:1011–1023
57. Siegbahn PEM, Crabtree RH (1997) J Am Chem Soc 119:3103–3113
58. Siegbahn PEM, Crabtree RH, Nordlund P (1998) J Biol Inorg Chem 3:314–317

59. Siegbahn PEM (1999) Inorg Chem 38:2880–2889
60. Siegbahn PEM, Blomberg MRA (1999) Ann Rev Phys Chem 50:221–249
61. Siegbahn PEM, Blomberg MRA (2000) Chem Rev 100:421–438
62. Blomberg MRA, Siegbahn PEM (2001) J Phys Chem B 105:9375–9386
63. Siegbahn PEM (2001) J Bio Inorg Chem 6:27–45
64. Siegbahn PEM, Blomberg MRA (2001) Theor Comp Chem 9:95–143
65. Blomberg MRA, Borowski T, Himo F, Liao RZ, Siegbahn PEM (2014) Chem Rev 114:3601–3658
66. Basch H, Mogi K, Musaev DG, Morokuma K (1999) J Am Chem Soc 121:7249−7256
67. Torrent M, Musaev DG, Morokuma K (2001) J Phys Chem B 105:322−327
68. Basch H, Musaev DG, Mogi K, Morokuma K (2001) J Phys Chem A 105:3615–3622
69. Torrent M, Musaev DG, Morokuma K, Basch H (2001) J Phys Chem B 105:4453–4463
70. Basch H, Musaev DG, Morokuma K (2001) J Phys Chem B105:8452–8460
71. Torrent M, Mogi K, Basch H, Musaev DG, Morokuma K (20001) J Phys Chem B 105:8616–8628
72. Torrent M, Musaev DG, Basch H, Morokuma K (2001) J Comp Chem 23:59–76
73. Torrent M, Vreven T, Musaev DG, Morokuma K, Farkas Ö, Schlegel HB (2002) J Am Chem Soc 124:192–193
74. Musaev DG, Basch H, Morokuma K (2002) J Am Chem Soc 124:4135–4148
75. Hoffmann M, Khavrutskii IV, Musaev DG, Morokuma K (2004) Int J Quant Chem 99:972–980
76. Dunietz BD, Beachy MD, Cao Y, Whittington DA, Lippard SJ, Friesner RA (2000) J Am Chem Soc 122:2828–2839
77. Gherman BF, Dunietz BD, Whittington DA, Lippard SJ, and Friesner RA (2001) J Am Chem Soc 123:3836−3837
78. Baik MH, Lee D, Friesner RA, Lippard SJ (2001) Israel J Chem 41:173−186
79. Friesner RA, Dunietz BD (2001) Acc Chem Res 34:351−358
80. Guallar V, Gherman BF, Miller WH, Lippard SJ, Friesner RA (2002) J Am Chem Soc 124:3377−3384
81. Ambundo EA, Friesner RA, Lippard SJ (2002) J Am Chem Soc 124:8770−8771
82. Guallar V, Gherman BF, Lippard SJ, Friesner RA (2002) Curr Opin Chem Biol 6:236−242
83. Baik MH, Gherman BF, Friesner RA, Lippard SJ (2002) J Am Chem Soc 124:14608−14615
84. Baik MH, Newcomb M, Friesner RA, Lippard SJ (2003) Chem Rev 103:2385−2420
85. Friesner RA, Baik MH, Gherman BF, Guallar V, Wirstam M, Murphy RB, Lippard SJ (2003) Coord Chem Rev 238−239:267−290
86. Gherman BF, Baik MH, Lippard SJ, Friesner RA (2004) J Am Chem Soc 126:2978−2990
87. Gherman BF, Lippard SJ, Friesner RA (2005) J Am Chem Soc 127:1025−1037
88. Rinaldo D, Philipp DM, Lippard SJ, Friesner RA (2007) J Am Chem Soc 129:3135−3147
89. Bochevarov AD, Friesner RA, Lippard SJ (2010) J Chem Theory Comput 6:3735–3749
90. Lovell T, Li J, Noodleman L (2001) Inorg Chem 40:5251–5266
91. Lovell T, Li J, Noodleman L (2001) Inorg Chem 40:5267–5278
92. Lovell T, Li J, Noodleman L (2002) J Biol Inorg Chem 7–8:799–809
93. Lovell T, Himo F, Han WG, Noodleman L (2003) Coord Chem Rev 238–239:211–232.
94. Lovell T, Han WG, Liu T, Noodleman L (2002) J Am Chem Soc 124:5890–5894
95. Noodleman L, Lovell T, Han WG, Li J, Himo F (2004) Chem Rev 104:459–508
96. Han WG, Noodleman L (2008) Inorg Chim Acta 361:973–986
97. Han WG, Noodleman L (2008) Inorg Chem 47:2975–2986
98. Saito T, Shoji M, Isobe H, Yamanaka S, Kitagawa Y, Yamada S, Kawakami T, Okumura M, Yamaguchi K (2010) Int J Quant Chem 110:2955–2981
99. Da Silva JCS, Pennifold RCR, Harvey JN, Rocha WR (2016) Dalton Trans 45:2492–2504
100. Balcells D, Clot E, Eisenstein O (2010) Chem Rev 110:749–823
101. Himo F (2017) J Am Chem Soc 139:6780–6786
102. Saillard JY, Hoffmann R (1984) J Am Chem Soc 106:2006–2026
103. Becke AD (1988) Phys Rev A 38:3098–3100
104. Becke AD (1993) J Chem Phys 98:5648–5652

105. Stephens PJ, Devlin FJ, Chabalowski CF, Frisch MJ (1994) J Phys Chem 98:11623–11627
106. Lee C, Yang W, Parr RG (1988) Phys Rev B 37:785–789
107. Vosko SH, Wilk L, Nusair M (1980) Can J Phys 58:1200–1211
108. Maseras F, Morokuma K (1995) J Comp Chem 16:1170–1179
109. Svensson M, Humel S, Froese RDJ, Matsuara T, Sieber S, Morokuma K (1996) J Phys Chem 100:19357–19363
110. Inorganic Chemistry, 4th ed. Shriver DF, Atkins PW, Overton TL, Rourke JP, Weller MT, Armstrong FA (2006) Oxford University Press, Oxford
111. Noodleman L (1981) J Chem Phys 74:5737–5743
112. Noodleman L, Davidson ER (1986) Chem Phys 109:131–143
113. Noodleman L, Peng CY, Case DA, Mouesca JM (1995) Coord Chem Rev 144:199–244
114. Yoshizawa K, Shiota Y (2006) J Am Chem Soc 128:9873−9881
115. Shiota Y, Yoshizawa K (2009) Inorg Chem 48:838−845
116. Shiota Y, Juhasz G, Yoshizawa K (2013) Inorg Chem 52:7907−7917
117. Itoyama S, Doitomi K, Kamachi T, Shiota Y, Yoshizawa K (2016) Inorg Chem 55:2771−2775
118. Schröder D, Schwarz H (1995) Angew Chem Int Ed 34:1973–1995
119. Schröder D, Schwarz H (1990) Angew Chem Int Ed 29:1433–1434
120. Schröder D, Fiedler A, Hrusak, J, Schwarz H (1992) J Am Chem Soc 114:1215–1222
121. Schröder D, Schwarz H, Clemmer DE, Chen Y, Armentrout PB, Baranov VI, Böhme DK (1997) Int J Mass Spectrom Ion Processes 161:175–191
122. Inoue T, Shiota Y, Yoshizawa K (2008) J Am Chem Soc 130:16890–16897
123. Chen PPY, Chan SI (2006) J Inorg Biochem 100:801–809
124. Chan SI, Chen KHC, Yu SSF, Chen CL, Kuo SSJ (2004) Biochemistry 43:4421–4430
125. Shaik S, Danovich D, Fiedler A, Schröder D, Schwarz H (1995) Helv Chim Acta 78:1393–1407
126. Fiedler A, Schröder D, Shaik S, Schwarz H (1994) J Am Chem Soc 116:10734–10741
127. Schröder D, Shaik S, Schwarz H (2000) Acc Chem Res 33:139–145
128. Shiota Y, Yoshizawa K (2003) J Chem Phys 118:5872–5879
129. Ravi M, Ranocchiari M, van Bokhoven JA (2017) Angew Chem Int Ed 56:16464–16483
130. BER Snyder ML Bols RA Schoonheydt BF Sels EI Solomon 2018 Chem Rev 118 2718 2768
131. Dinh KT, Sullivan MM, Serna P, Meyer RJ, Dinca M, Roman-Leshkov Y (2018) ACS Catal 8:8306 -8313
132. Snyder BER, Vanelderen P, Schoonheydt RA, Sels, BF, Solomon EI (2018) J Am Chem Soc 140:9236–9243
133. Yumura T, Takeuchi M, Kobayashi H, Kuroda Y (2009) Inorg Chem 48:508–517
134. Yumura T, Hirose Y, Wakasugi T, Kuroda Y, Kobayashi H (2016) ACS Catal 6:2487–2495
135. Castillo R. G., Banerjee R., Allpress C. J., Rohde G. T., Bill E., Que, Jr. L., Lipscomb J. D., and DeBeer S., (2017) J. Am. Chem. Soc. 139: 18024-18033
136. Cutsail III G. E., Banerjee R., Zhou A., Que, Jr. L., Lipscomb J. D., and DeBeer S., (2018) J. Am. Chem. Soc., 140: 16807-16820.
137. Banerjee R., Proshylyakov Y., Lipscomb J. D., and Proshylyakov D. A., (2015) Nature, 518: 431-434.
138. Ross M. O., MacMillan F., Wang J., Nisthal A., Lawton T. J., Olafson B. D., Mayo S. L., Rosenzweig A. C., Hoffman B. M., (2019) Science, 364: 566-570.

Mechanistic Understanding of Methane Hydroxylation by Cu-Exchanged Zeolites

Muhammad Haris Mahyuddin, Hermawan Kresno Dipojono, and Kazunari Yoshizawa

Abstract Methane direct conversion to methanol or methane hydroxylation by Cu-exchanged zeolite catalysts through a stepwise approach is a promising process for fulfilling the increasing energy demand in the midst of declining fossil energy and unreadiness of renewable energy technologies. However, there are still many challenges hampering this process from yielding methanol with amount that meets industrial expectations. Therefore, understanding the mechanism of how the strong C–H bond of methane is cleaved and how methanol is formed only in the presence of a particular solvent could help optimizing the process or finding novel approaches. In this chapter, we overview the mechanistic understanding on the methane's C–H cleavage through the homolytic hydrogen atom abstraction and evaluate the formation of methanol through a mechanism where a surface methoxy is formed as an intermediate species and a water molecule acts as a reactant or an assistant to form methanol through a hydrolysis or an indirect HO–CH_3 rebound.

Keywords Copper · Zeolites · Methane activation · Methanol formation

1 Introduction

Nature, through a certain kind of enzymes that contain iron or copper centers in its active site (soluble or particulate methane monooxygenase, sMMO or pMMO, respectively), has shown its ability to hydroxylate methane with O_2 under ambient conditions [1–4]. Since the reaction between the triplet (O_2) and singlet (CH_4)

M. H. Mahyuddin (✉) · H. K. Dipojono
Research Group of Advanced Functional Materials, Faculty of Industrial Technology, Institut Teknologi Bandung, Bandung 40132, Indonesia
e-mail: haris@tf.itb.ac.id

H. K. Dipojono
e-mail: dipojono@tf.itb.ac.id

K. Yoshizawa
Institute for Materials Chemistry and Engineering, Kyushu University, 744 Motooka Nishi-ku, Fukuoka 819-0395, Japan
e-mail: kazunari@ms.ifoc.kyushu-u.ac.jp

K. Yoshizawa (ed.), *Direct Hydroxylation of Methane*,
https://doi.org/10.1007/978-981-15-6986-9_4

substrates is known to be spin-forbidden, it is reasonable that the metal centers in MMOs facilitate a spin inversion for O_2, resulting in a metal-superoxo ($M^{n+1}O_2^{\cdot -}$), -peroxo ($M^{n+2}O_2^{2-}$), or -oxo ($M^{n+4}[O^{2-}]_2$) species that is favorable for the oxidation of methane. However, controlling active sites in enzymes is extremely difficult, especially when considering mass production. Therefore, comparable metal active sites isolated in rigid lattices of zeolite have been developed as synthetic catalysts for methane hydroxylation at low temperature. Zeolites are micropores crystalline minerals composed of aluminate, silicate, and countercations acting as Brønsted or Lewis acid sites. These countercations are not covalently bound to the zeolite framework and thus can readily be substituted with a variety of metal cations through conventional aqueous ion-exchange methods to form reduced metal centers.

Methane hydroxylation by metal-exchanged zeolites is in general a stepwise (stoichiometric looping) process although the continuous (catalytic) process is also available but disfavored due to its low methanol yield and selectivity. As shown in Fig. 1, the stepwise process involves three separate steps including (1) activation of the metal–zeolite catalyst by an oxidant at high temperature, (2) methane reaction at temperatures below 200 °C, and (3) methanol extraction using a solvent or steam at 25–200 °C. A number of successful experiments have been reported for Fe- and Cu-exchanged zeolites. Panov and co-workers are the pioneers who discovered a highly reactive α-oxygen site upon N_2O decomposition on Fe-ZSM-5 [5–7]. This catalyst activates methane at room temperature and produces methanol with 80% selectivity. The active site has been recently found to be a mononuclear $[FeO]^{2+}$ species anchored on the six-membered ring (6-MR) of zeolites [8, 9]. Besides N_2O, hydrogen peroxide was also reported to activate Fe-ZSM-5 and form a different

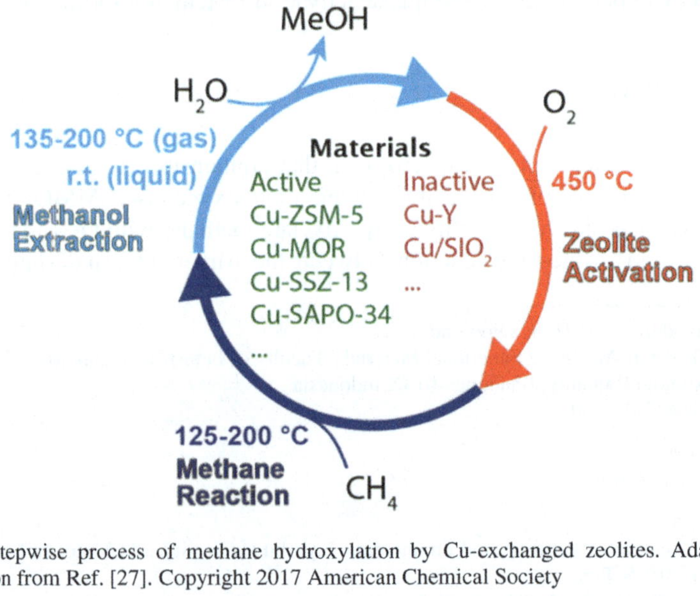

Fig. 1 Stepwise process of methane hydroxylation by Cu-exchanged zeolites. Adapted with permission from Ref. [27]. Copyright 2017 American Chemical Society

active site structure that catalyzes methane hydroxylation in liquid phase [10–13]. Following the work of Panov and co-workers, Schoonheydt and co-workers reported O_2-activated Cu-ZSM-5 and Cu-MOR zeolites, which oxidize methane at 125 °C to yield methanol with 98% selectivity [14, 15]. A significant amount of studies on these particular Cu–zeolite catalysts have been reported, owing particularly to their ability to activate a variety of oxidants including O_2 [14], N_2O [14], H_2O_2 [10], and H_2O [16]. The $[Cu_2(\mu\text{-}O)]^{2+}$ and $[Cu_3(\mu\text{-}O)_3]^{2+}$ species have been proposed as the active sites and debated for the past ten years [17–19], but no consensus has been reached as to its structure. Beside iron and copper, Co, Ni, Zn, and Rh sites in zeolites were also reported to be active for methane hydroxylation at low temperature [20–26].

It is known that the oxo/oxyl species of the Fe and Cu active sites in zeolites abstract one H atom of methane through three possible mechanisms, namely heterolytic, homolytic, and Fenton-type H-atom abstraction (HAA) mechanisms [28, 29]. In the heterolytic HAA mechanism (Scheme 1, pathway 1), the C–H bond is activated by the metal–oxo pair, where the metal and the oxo behave as acid and base, respectively. Through a four-center transition state, a methyl ligand coordinated to the metal is then formed while the abstracted H atom is accepted by the oxo. Note that a C–H bond cleavage with this mechanism is not accompanied by any redox processes within the active site [29]. The oxidation of the resultant methyl moiety takes place in the subsequent formation of methanol, which is usually the rate-determining step [30]. In the homolytic mechanism (Scheme 1, pathway 2), the C–H bond cleavage of methane results in two radical species, namely a methyl radical and a formally H radical, during the transition state [29]. The H radical is readily reduced with one electron to form an O–H bond while the methyl radical stays uncoordinated. The subsequent direct $HO–CH_3$ rebound forming methanol then completes the two-electron oxidation process. In contrast to the rebound step in the heterolytic mechanism, that in this mechanism is usually low barrier [31, 32], thus making the C–H bond cleavage the rate-determining step. The Fenton-type mechanism (Scheme 1, pathway 3) is actually similar to the radical-involving mechanism, but here an $^{\cdot}OH$ radical is antecedently formed from H_2O_2 decomposition and acts

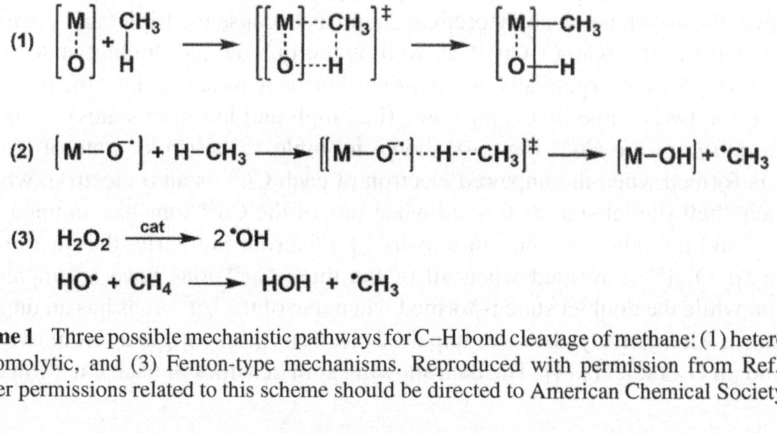

Scheme 1 Three possible mechanistic pathways for C–H bond cleavage of methane: (1) heterolytic, (2) homolytic, and (3) Fenton-type mechanisms. Reproduced with permission from Ref. [29]. Further permissions related to this scheme should be directed to American Chemical Society

as the active species abstracting the H atom of methane. For further discussion on methane hydroxylation by metal-exchanged zeolites, interested readers are invited to read recent excellent reviews [33–39].

Although most of theoretical works predict that methanol is formed through a direct HO–CH_3 recombination, several experimental works suggested that a surface methoxy is first formed as an intermediate state and then followed by water hydrolysis to form methanol [40, 41]. However, the fact that water acts only as a solvent and the catalysts need to be reactivated after each cycle of the reaction strongly indicates that the CH_3 species recombine with the OH originated from the active site, not the OH originated from water splitting. Despite this, theoretical understanding would be worthful to evaluate all possible reaction routes for methanol formation. In this chapter, we overview recent advances in the mechanistic understanding of the homolytic C–H bond cleavage of methane by the $[Cu–O–Cu]^{2+}$ and $[Cu_3(\mu\text{-}O)_3]^{2+}$ active species in zeolites and deliver our analysis on the mechanism of methanol formation by performing density functional theory (DFT) calculations. We present energy diagrams for the formation of methanol through two possible reaction routes, where the formation of a surface methoxy and the adsorption of a water molecule are involved.

2 Methane Hydroxylation by $[Cu_2(\mu\text{-}O)]^{2+}$ and $[Cu_3(\mu\text{-}O)_3]^{2+}$ in Zeolites

2.1 Mechanism of C–H Activation

The C–H activation of methane by $[Cu–O–Cu]^{2+}$ and $[Cu_3(\mu\text{-}O)_3]^{2+}$ active species in zeolites mostly reported in theoretical works follows the homolytic HAA mechanism (Scheme 1, pathway 2) [17, 19, 42–44]. Here, the Cu^{2+} ions provide electrons to the bridging O atoms(s), resulting in a radical O^- that is highly reactive toward CH_4. The ground states of $[Cu–O–Cu]^{2+}$ and $[Cu_3(\mu\text{-}O)_3]^{2+}$ active species have been an intensive discussion between theoretical chemists because the triplet and open-shell singlet states in the $[Cu–O–Cu]^{2+}$ as well as the quartet and doublet states in the $[Cu_3(\mu\text{-}O)_3]^{2+}$ are energetically competitive. This is reasonable since the difference between the two competitive spin states (i.e., high and low spin states) is only the spin direction of the Cu^{2+} ions. As shown in Table 1, the triplet state of $[Cu–O–Cu]^{2+}$ is formed when the unpaired electron of each Cu^{2+} is an α electron, whereas the open-shell singlet state is formed when one of the Cu^{2+} ions has an unpaired α electron and the other one has an unpaired β electron. Similarly, the quartet state of $[Cu_3(\mu\text{-}O)_3]^{2+}$ is formed when all of the three Cu^{2+} ions have an unpaired α electron while the doublet state is formed when one of the Cu^{2+} ions has an unpaired β electron. Alternatively, the three unpaired electrons might originate from two Cu^{2+} ions and an O^- radical [44]. The data showing spin densities of each Cu^{2+} ion in the

Table 1 Possible spin configurations in the $[Cu_2(\mu\text{-}O)]^{2+}$ and $[Cu_3(\mu\text{-}O)_3]^{2+}$ active sites[a]

Spin state	$[Cu_2(\mu\text{-}O)]^{2+}$	$[Cu_3(\mu\text{-}O)_3]^{2+}$
High-spin configuration	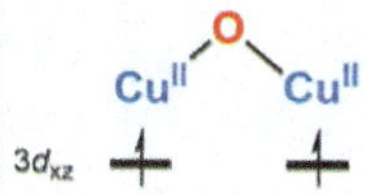 Triplet state	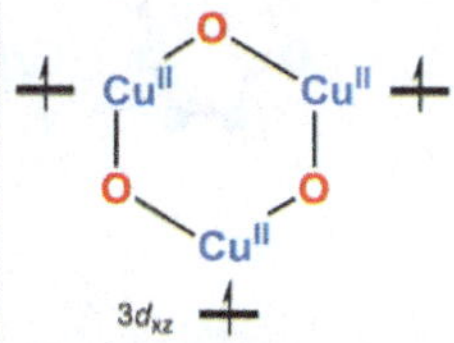 Quartet state
Low-spin configuration	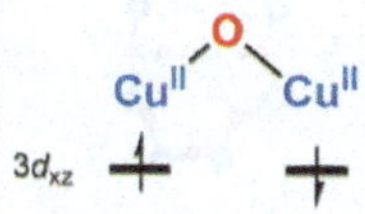 Open-shell singlet state	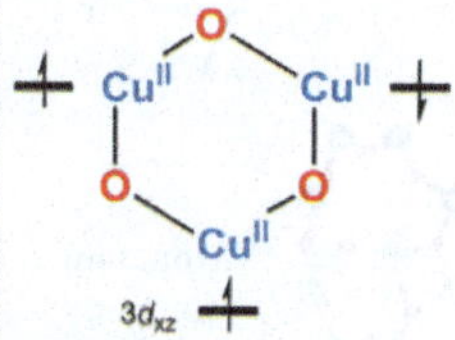 Doublet state

[a]Only the Cu 3d orbital with an unpaired electron is shown for simplicity

$[Cu–O–Cu]^{2+}$ and $[Cu_3(\mu\text{-}O)_3]^{2+}$ species has been presented in our recent work [43, 44].

In the homolytic HAA mechanism, one H atom of methane is abstracted by the $[Cu–O–Cu]^{2+}$ and $[Cu_3(\mu\text{-}O)_3]^{2+}$ active sites to form an intermediate structure (referred to as radical intermediate), which involves a $^{\cdot}CH_3$ radical species and an O–H moiety and is less stable than the reactant complex (CH_4 + catalyst, see Fig. 2). It has been theoretically predicted that the strength of the formed O–H bond significantly determines the stability of radical intermediate and thus the nature of the transition state [45]. Thus, one can predict the activation barrier of methane's C–H bond cleavage from the hydrogen affinity of the active O atom; the stronger hydrogen affinity is, the lower the activation barrier is [45, 46]. In addition, it is essential for the bridging O atom(s) of the active species to have a radical character (i.e., spin density approaching one) because only such an unstable species can be highly reactive toward methane [47]. However, the O-atom spin density indeed cannot be used as a descriptor for the reactivity of any trivial catalysts as it does not create a linear relationship with the reactivity of known catalysts [46].

The discussion of spin state change/inversion during the C–H bond cleavage of methane is also a hot issue among theoretical chemists. Formally, when the H atom is cleaved by the $[Cu^{2+}–O^{2-}–Cu^{2+}]^{2+}$ active site, for example, and then approaches the bridging O atom, it forms a hydroxide $[OH]^-$ ion that results in a one-electron reduction on one of the Cu^{2+} centers and a $^{\cdot}CH_3$ radical that keeps the spin state in the triplet or open-shell singlet state, but not change to the closed-shell singlet state (Scheme 2a). Similarly, when one considers $[Cu^{+}–O^{\cdot-}–Cu^{2+}]^{2+}$ as the ground-state electronic structure (see Scheme 2b), the formed radical intermediate would involve a neutral hydroxyl ($^{\cdot}OH$) instead of $[OH]^-$, resulting in a one-electron reduction on the Cu^{2+} center in order to keep the total charge of the active site + 2, but the ground

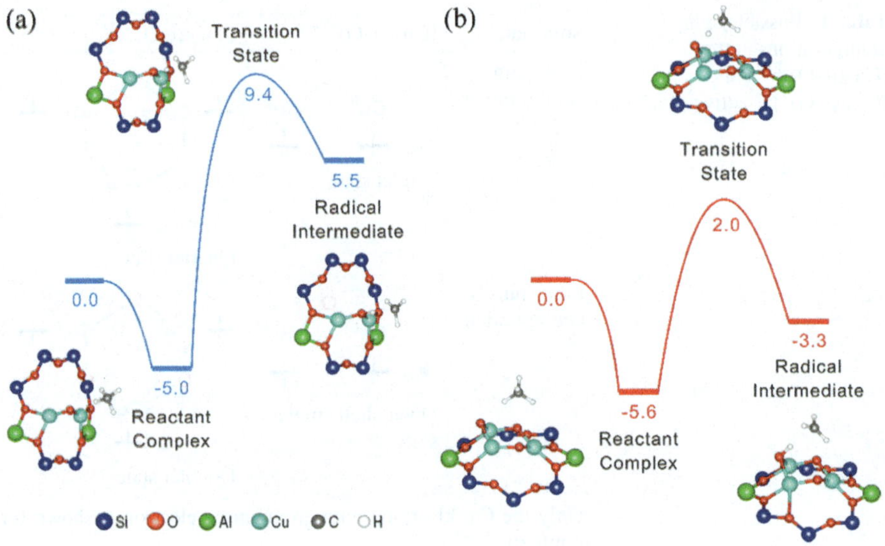

Fig. 2 Homolytic hydrogen atom abstraction of methane over **a** $[Cu_2(\mu\text{-}O)]^{2+}$-MOR and **b** $[Cu_3(\mu\text{-}O)_3]^{2+}$-MOR in the corresponding ground state, calculated by using periodic DFT-PBE method. Energies are in kcal/mol. Reproduced with permission from Ref. [43]. Copyright 2018 American Chemical Society

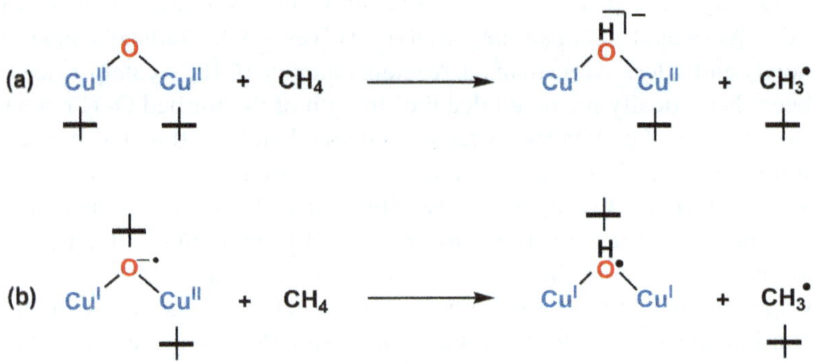

Scheme 2 Changes in the electronic structure of $[Cu–O–Cu]^{2+}$ active species along the reaction of methane activation (reactant complex → radical intermediate). The orbital with an unpaired electron is shown for Cu^{2+}, $^{\cdot}CH_3$, $O^{\cdot}$, and $^{\cdot}OH$

state remains unchanged (i.e., triplet or open-shell singlet state) since each of the hydroxyl and the $^{\cdot}CH_3$ has one unpaired electron (Scheme 2b).

2.2 *Mechanism of CH_3OH Formation*

There has been a number of mechanisms proposed for elucidating the formation of methanol after the methane activation. One of the most well-accepted mechanisms is the direct rebound/recombination between the $^{\cdot}CH_3$ radical and the OH moiety (Scheme 3a). This mechanism was proposed due to the fact that the resonance Raman (rR) spectroscopic peak assigned to the bridging O atom of the $[Cu–O–Cu]^{2+}$ disappears when methane is reacted to form methanol [17], indicating that this O atom forms a CH_3OH molecule coordinated to the Cu centers. Afterward, water is used to assist the methanol desorption, making the release of methanol exothermic [43]. The resultant H_2O bound to the Cu centers might be the reason why the catalyst needs to be dehydrated and reactivated at high temperature after each cycle of reaction [48]. However, Alayon et al. [40] and Dyballa et al. [41] recently suggested a surface methoxy (CH_3O–) species as a key intermediate structure that subsequently forms methanol in the presence of water (Scheme 3b).

Li et al. [42] reported energy diagrams showing that the formation of a surface methoxy on either $[Cu–O–Cu]^{2+}$ or $[Cu_3(\mu\text{-}O)_3]^{2+}$ active site is energetically more preferable than the formation of methanol through the direct radical recombination mechanism (Fig. 3) [42]. This indicates that the methoxy species may actually be the next intermediate structure before a CH_3OH molecule is formed through a hydrolysis process. Consequently, to rationalize the fact that the bridging O atom disappears from the system, the bridging OH moiety should bind the H atom released from the hydrolysis process and then form an uncoordinated H_2O molecule (Scheme 3b(i)). Alternatively, the incoming H_2O molecule replaces the bridging OH moiety to bind the Cu centers while the CH_3 and OH recombine with each other to form methanol (Scheme 3b(ii), referred to as an indirect HO–CH_3 rebound).

To evaluate the hypothesis described above, we perform DFT calculations using the same computational methods as those in Ref. [43]. Figure 4 shows the calculation results, where the data of C–H cleavage in the triplet state is taken from our previous work [43] while the rest is originally calculated in the closed-shell

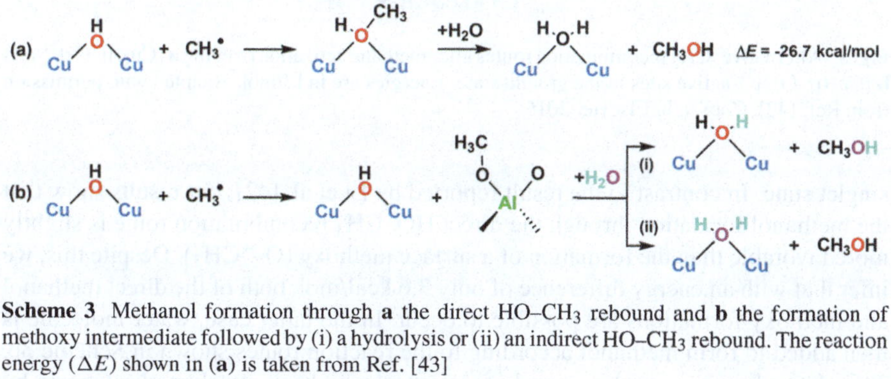

Scheme 3 Methanol formation through **a** the direct HO–CH_3 rebound and **b** the formation of methoxy intermediate followed by (i) a hydrolysis or (ii) an indirect HO–CH_3 rebound. The reaction energy (ΔE) shown in (**a**) is taken from Ref. [43]

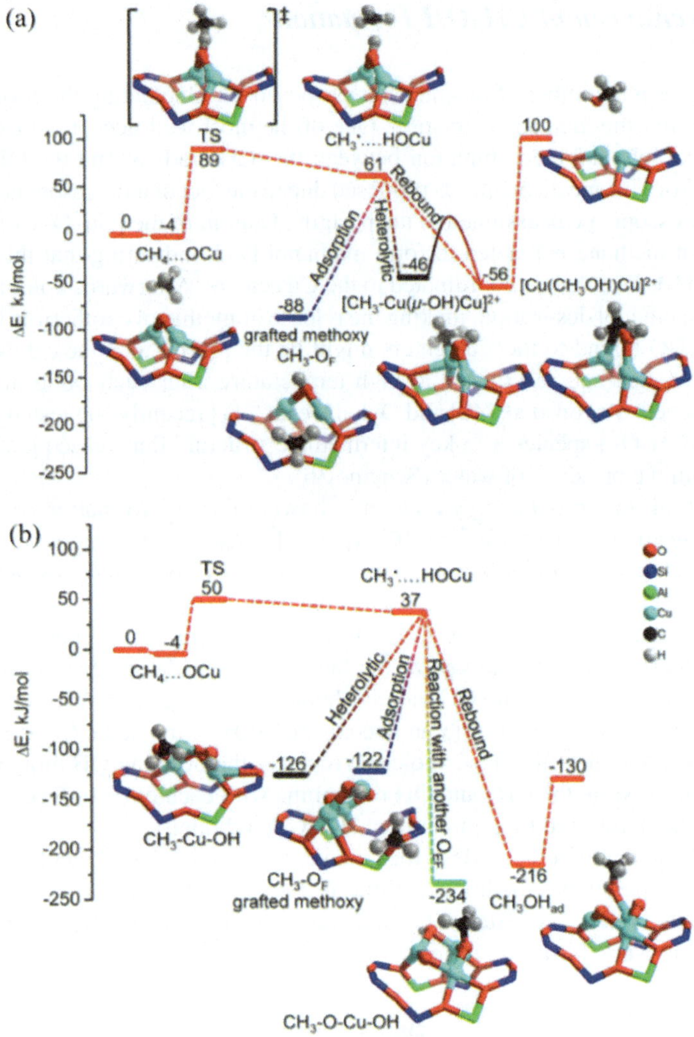

Fig. 3 Alternative $^{\cdot}CH_3$ recombination routes after methane activation over the **a** $[Cu_2(\mu\text{-}O)]^{2+}$ and **b** $[Cu_3(\mu\text{-}O)_3]^{2+}$ active sites in the ground state. Energies are in kJ/mol. Adapted with permission from Ref. [42]. Copyright Elsevier 2016

singlet state. In contrast to the result reported by Li et al. [42], our results show that the methanol formation through the direct HO–CH_3 recombination route is slightly more favorable than the formation of a surface methoxy (O–$^{+}CH_3$). Despite this, we infer that with an energy difference of only 3.6 kcal/mol, both of the direct methanol and methoxy formations are possible to occur. In the latter case, water molecule is then added to form methanol according to the reaction routes shown in Scheme 3b. Two H_2O adsorption modes, namely front-attack and back-attack modes (shown by

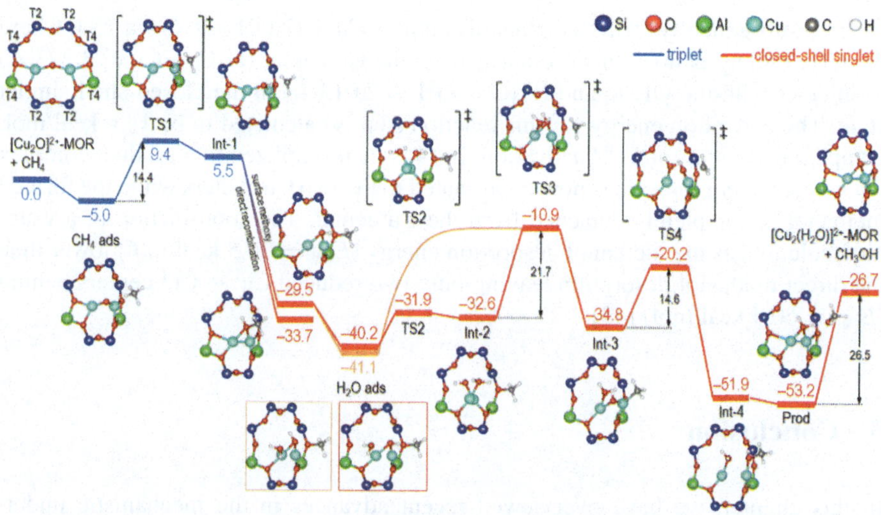

Fig. 4 Calculated energy diagrams of methane's C–H cleavage and methanol formation through a surface methoxy intermediate over $[Cu_2(\mu\text{-}O)]^{2+}$-MOR (Al pair at the T4/T4 site). Energies are in kcal/mol. TS and Int stand for transition state and intermediate, respectively. Data of the C–H cleavage adapted with permission from Ref. [43]. Copyright 2018 American Chemical Society

orange and red lines, respectively), are considered for the indirect HO–CH_3 rebound (Scheme 3b(ii)) and hydrolysis (Scheme 3b(i)) routes, respectively.

The front-attack H_2O adsorption, however, leads directly to the formation of a methyl complex Cu–CH_3 (Int-3) via TS3 and requires a quite high activation barrier of 30.2 kcal/mol. On the contrary, the back-attack H_2O adsorption, which is only by 0.9 kcal/mol less stable than the front-attack H_2O adsorption, leads first to the formation of an OH moiety from water (Int-2) before forming Int-3. Although this reaction step via TS2 requires a low activation barrier (9.2 kcal/mol), the existence of Int-2 is questionable since Int-2 is energetically less stable than Int-3. Assuming that Int-2 really exist as an intermediate species, the activation barrier for forming Int-3 is calculated to be 21.7 kcal/mol. Regardless of which route (i.e., indirect HO–CH_3 rebound or hydrolysis) being actually more favorable, both activation energy values are higher than the activation barrier for the C–H cleavage (14.4 kcal/mol). This, interestingly, is consistent with the fact that at the same reaction temperature (e.g., 200 °C), the methanol extraction procedure using water steam takes longer time than the methane activation and is performed several times to obtain maximum yield of methanol [16, 49, 50]. Furthermore, at lower temperature, when water liquid is used for the extraction, the amount of methanol extracted becomes lower significantly [49, 51]. In TS3, the CH_3 is detached from the zeolite framework with a $H_3C \cdots O$ distance (2.135 Å) being moderate and thus allowing the ground state to remain at the closed-shell singlet state.

Int-3 subsequently forms a methanol complex Cu–CH_3OH (Int-4) via TS4, where the CH_3 moiety is now uncoordinated from the Cu center ($H_3C \cdots Cu = 2.223$ Å) to interact with the OH ligand ($H_3C \cdots OH = 2.013$ Å) in the closed-shell singlet state. The activation energy for this reaction step is calculated to be 14.6 kcal/mol, comparable to that for C–H activation. Int-4 is then stabilized to a product complex where the water molecule is now coordinated to the two Cu centers while the formed methanol is completely detached from the Cu center. The coordination of a water molecule makes the methanol desorption energy ($E_{des} = 26.5$ kcal/mol) lower than the direct methanol desorption leaving only two reduced $Cu^I \cdots Cu^I$ centers behind ($E_{des} = 60.4$ kcal/mol) [43].

3 Conclusion

In this chapter, we have overviewed recent advances in the mechanistic understanding of methane's C–H bond cleavage by the $[Cu–O–Cu]^{2+}$ and $[Cu_3(\mu\text{-}O)_3]^{2+}$ active species in zeolites. Being the most well-accepted mechanism, the homolytic hydrogen atom transfer mechanism becomes our focus of discussion. We show how the high spin state and low spin state of the two active species are highly competitive in energetics since the difference between the two spin states lies only on the spin direction of the Cu^{2+} ions. In addition, we also demonstrate how the ground state along the cleavage process on the $[Cu–O–Cu]^{2+}$ active site (triplet or open-shell singlet state) formally remains unchanged.

More importantly, here we have reviewed the mechanism of methanol formation from the cleaved $^{\cdot}CH_3$ radical and the formed OH moiety, which has been debated in literatures. While most of theoretical works suggested the direct HO–CH_3 rebound as the most probable mechanism, recent experimental works suggested that the $^{\cdot}CH_3$ radical first forms a surface methoxy species (O_F–CH_3, where O_F is an O atom from the zeolite framework) as an intermediate state before the formation of methanol. Here, we show that the formation of O_F–CH_3 species is as energetically probable as the direct formation of methanol. The addition of a water molecule on the methoxy intermediate facilitates the formation of a methyl complex (Cu–CH_3) via two possible pathways that require activation barriers of 21.7 and 30.2 kcal/mol, both of which are higher than that for cleaving the C–H bond of methane (14.4 kcal/mol). This, however, is reasonable when one considers the fact that at the same reaction temperature, the methanol extraction procedure using water steam takes longer time than the methane activation and is performed several times to obtain maximum yield of methanol [16, 49, 50].

Acknowledgements M.H.M. and H.K.D. acknowledge a research fund from the Indonesian Ministry of Research and Technology/National Agency for Research and Innovation, and Indonesian Ministry of Education and Culture under World Class University (WCU) Program managed by Institut Teknologi Bandung. K.Y. acknowledges the KAKENHI with grant numbers of JP24109014, JP15K13710, and JP17H03117 from Japan Society for the Promotion of Science (JSPS) and the Ministry of Education, Culture, Sports, Science and Technology of Japan (MEXT), the MEXT

Projects of "Integrated Research Consortium on Chemical Sciences" and "Elements Strategy Initiative to Form Core Research Center," and the JST-CREST with grant numbers of JPMJCR15P5 and JPMJCR18R2. The computations were performed at the Research Institute for Information Technology (Kyushu University).

References

1. Rosenzweig AC, Frederick CA, Lippard SJ, Nordlund P (1993) Nature 366:537–543
2. Lieberman RL, Rosenzweig AC (2005) Nature 434:177–182
3. Chan SI, Chen KH-C, Yu SS-F, Chen C-L, Kuo SS-J (2004) Biochemistry 43:4421–4430
4. Mahyuddin MH, Shiota Y, Staykov A, Yoshizawa K (2018) Acc Chem Res 51:2382–2390
5. Panov GI, Sobolev VI, Kharitonov AS (1990) J Mol Catal 61:85–97
6. Sobolev VI, Dubkov KA, Panna OV, Panov GI (1995) Catal Today 24:251–252
7. Panov GI, Sobolev VI, Dubkov KA, Parmon VN, Ovanesyan NS, Shilov AE, Shteinman AA (1997) React Kinet Catal Lett 61:251–258
8. Snyder BER, Vanelderen P, Bols ML, Hallaert SD, Böttger LH, Ungur L, Pierloot K, Schoonheydt RA, Sels BF, Solomon EI (2016) Nature 536:317–321
9. Bols ML, Hallaert SD, Snyder BER, Devos J, Plessers D, Rhoda HM, Dusselier M, Schoonheydt RA, Pierloot K, Solomon EI, Sels BF (2018) J Am Chem Soc 140:12021–12032
10. Hammond C, Forde MM, Ab Rahim MH, Thetford A, He Q, Jenkins RL, Dimitratos N, Lopez-Sanchez JA, Dummer NF, Murphy DM, Carley AF, Taylor SH, Willock DJ, Stangland EE, Kang J, Hagen H, Kiely CJ, Hutchings GJ (2012) Angew Chem Int Ed 51:5129–5133
11. Hammond C, Dimitratos N, Lopez-Sanchez JA, Jenkins RL, Whiting G, Kondrat SA, ab Rahim MH, Forde MM, Thetford A, Hagen H, Stangland EE, Moulijn JM, Taylor SH, Willock DJ, Hutchings GJ (2013) ACS Catal 3:1835–1844
12. Hammond C, Hermans I, Dimitratos N (2015) ChemCatChem 7:434–440
13. Xiao P, Wang Y, Nishitoba T, Kondo JN, Yokoi T (2019) Chem Commun 55:2896–2899
14. Groothaert MH, Smeets PJ, Sels BF, Jacobs PA, Schoonheydt RA (2005) J Am Chem Soc 127:1394–1395
15. Smeets PJ, Groothaert MH, Schoonheydt RA (2005) Catal Today 110:303–309
16. Sushkevich VL, Palagin D, Ranocchiari M, van Bokhoven JA (2017) Science 356:523–527
17. Woertink JS, Smeets PJ, Groothaert MH, Vance MA, Sels BF, Schoonheydt RA, Solomon EI (2009) Proc Natl Acad Sci U S A 106:18908–18913
18. Vanelderen P, Snyder BER, Tsai M-L, Hadt RG, Vancauwenbergh J, Coussens O, Schoonheydt RA, Sels BF, Solomon EI (2015) J Am Chem Soc 137:6383–6392
19. Grundner S, Markovits MAC, Li G, Tromp M, Pidko EA, Hensen EJM, Jentys A, Sanchez-Sanchez M, Lercher JA (2015) Nat Commun 6:7546
20. Beznis N V., Weckhuysen BM, Bitter JH (2010) Catal Letters 136:52–56
21. Shan J, Huang W, Nguyen L, Yu Y, Zhang S, Li Y, Frenkel AI, Tao F (2014) Langmuir 30:8558–8569
22. Mahyuddin MH, Yoshizawa K (2018) Catal Sci Technol 8:5875–5885
23. Oda A, Ohkubo T, Yumura T, Kobayashi H, Kuroda Y (2019) Inorg Chem 58:327–338
24. Shan J, Li M, Allard LF, Lee S, Flytzani-Stephanopoulos M (2017) Nature 551:605–608
25. Mahyuddin MH, Tanaka S, Shiota Y, Yoshizawa K (2020) Bull Chem Soc Jpn 93: 345–354
26. Shah MA, Raynes S, Apperley DC, Taylor RA (2020) ChemPhysChem doi: 10.1002/cphc.201900973
27. Tomkins P, Ranocchiari M, van Bokhoven JA (2017) Acc Chem Res 50:418–425
28. Olivos-Suarez AI, Szécsényi À, Hensen EJM, Ruiz-Martinez J, Pidko EA, Gascon J (2016) ACS Catal 6:2965–2981
29. Szécsényi Á, Li G, Gascon J, Pidko EA (2018) ACS Catal 8:7961–7972

30. Mahyuddin MH, Staykov A, Shiota Y, Yoshizawa K (2016) ACS Catal 6:8321–8331
31. Göltl F, Michel C, Andrikopoulos PC, Love AM, Hafner J, Hermans I, Sautet P (2016) ACS Catal 6:8404–8409
32. Mahyuddin MH, Staykov A, Shiota Y, Miyanishi M, Yoshizawa K (2017) ACS Catal 7:3741–3751
33. Snyder BER, Bols ML, Schoonheydt RA, Sels BF, Solomon EI (2018) Chem Rev 118:2718–2768
34. Kulkarni AR, Zhao Z-J, Siahrostami S, Nørskov JK, Studt F (2018) Catal Sci Technol 8:114–123
35. Ravi M, Ranocchiari M, van Bokhoven JA (2017) Angew Chem Int Ed 56:16464–16483
36. Dinh KT, Sullivan MM, Serna P, Meyer RJ, Dincă M, Román-Leshkov Y (2018) ACS Catal 8:8306–8313
37. Borfecchia E, Beato P, Svelle S, Olsbye U, Lamberti C, Bordiga S (2018) Chem Soc Rev 47:8097–8133
38. Li G, Pidko EA (2019) ChemCatChem 11:134–156
39. Mahyuddin MH, Shiota Y, Yoshizawa K (2019) Catal Sci Technol 9:1744–1768
40. Alayon EMC, Nachtegaal M, Bodi A, van Bokhoven JA (2014) ACS Catal 4:16–22
41. Dyballa M, Thorshaug K, Pappas DK, Borfecchia E, Kvande K, Bordiga S, Berlier G, Lazzarini A, Olsbye U, Beato P, Svelle S, Arstad B (2019) ChemCatChem 11:5022–5026
42. Li G, Vassilev P, Sanchez-Sanchez M, Lercher JA, Hensen EJM, Pidko EA (2016) J Catal 338:305–312
43. Mahyuddin MH, Tanaka T, Shiota Y, Staykov A, Yoshizawa K (2018) ACS Catal 8:1500–1509
44. Mahyuddin MH, Staykov A, Saputro AG, Agusta MK, Dipojono HK, Yoshizawa K (2020) J Phys Chem C 124:18112–18125
45. Latimer AA, Kulkarni AR, Aljama H, Montoya JH, Yoo JS, Tsai C, Abild-Pedersen F, Studt F, Nørskov JK (2017) Nat Mater 16:225–229
46. Liu C, Li G, Pidko EA (2018) Small Methods 2:1800266
47. Dietl N, Schlangen M, Schwarz H (2012) Angew Chem Int Ed 51:5544–5555
48. Ikuno T, Grundner S, Jentys A, Li G, Pidko E, Fulton J, Sanchez-Sanchez M, Lercher JA (2019) J Phys Chem C 123:8759–8769
49. Wulfers MJ, Lobo RF, Ipek B, Teketel S (2015) Chem Commun 51:4447–4450
50. Park MB, Ahn SH, Ranocchiari M, van Bokhoven J (2017) ChemCatChem 9:3705–3713
51. Sheppard T, Hamill CD, Goguet A, Rooney DW, Thompson JM (2014) Chem Commun 50:11053–5

Oxidative Activation of Metal-Exchanged Zeolite Catalysts for Methane Hydroxylation

Muhammad Haris Mahyuddin

Abstract The stepwise approach of methane direct conversion to methanol by Fe- and Cu-exchanged zeolite catalysts involves a pre-activation of the catalysts by an oxidant at high temperature. Since an isothermal process of this approach at temperature below 200 °C is highly desirable, understanding the mechanism of the oxidative activation is important for developing strategies to reduce the activation temperature. In this chapter, I provide an overview of recent theoretical understanding in the mechanisms of the Fe- and Cu-oxo active site formations from N_2O, H_2O_2, and O_2 activations. Density functional theory calculations show that N_2–O, HO–OH, and O=O cleavages on the reduced Fe and Cu centers can actually be achieved with low activation barriers. However, as also suggested by recent experimental works, high temperature is required to form the precursors, i.e. the reduced metal centers. I conclude this overview with forward looking aspects that should be addressed in the future.

Keywords Catalyst activation by oxidant · Active site formation · Iron · Copper · Zeolites

1 Introduction

As also described in Chap. 4, methane hydroxylation by metal-exchanged zeolites is in general a stepwise (stoichiometric looping) process although the continuous (catalytic) process is also available but disfavored due to its low methanol yield and selectivity [1, 2]. As shown in Fig. 1, the stepwise process involves three separate steps including (1) activation of the metal–zeolite catalyst by an oxidant at high temperature, (2) methane reaction at temperatures below 200 °C, and (3) methanol extraction using a solvent or steam at 25–200 °C. A number of successful experiments have been reported for Fe- and Cu-exchanged zeolites. Panov and co-workers

M. H. Mahyuddin (✉)
Research Group of Advanced Functional Materials, Faculty of Industrial Technology,
Institut Teknologi Bandung, Bandung 40132, Indonesia
e-mail: haris@tf.itb.ac.id

K. Yoshizawa (ed.), *Direct Hydroxylation of Methane*,
https://doi.org/10.1007/978-981-15-6986-9_5

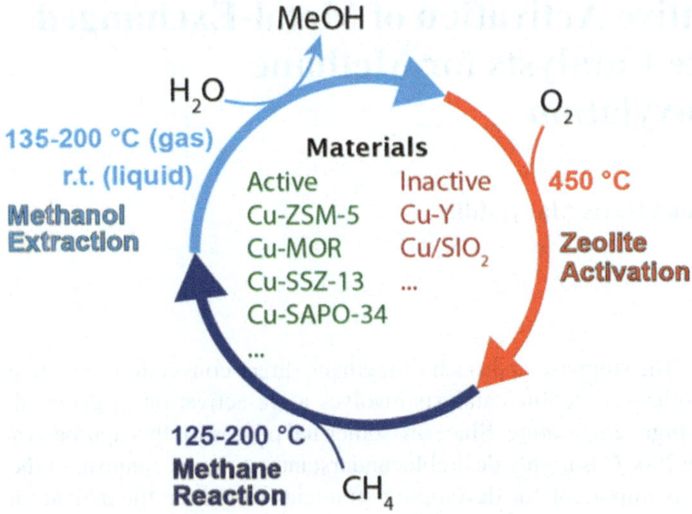

Fig. 1 Stepwise process of methane hydroxylation by Cu-exchanged zeolites. Adapted with permission from Ref. [22]. Copyright 2017 American Chemical Society

are the pioneers who discovered a highly reactive α-oxygen site upon N_2O decomposition on Fe-ZSM-5 [3–5]. This catalyst activates methane at room temperature and produces methanol with 80% selectivity. The active site has been recently found to be a mononuclear $[FeO]^{2+}$ species anchored on the six-membered ring (6-MR) of zeolites [6, 7]. Besides N_2O, hydrogen peroxide was also reported to activate Fe-ZSM-5 to form a different active site structure that catalyzes methane hydroxylation in liquid phase [8–11]. Following the work of Panov and co-workers, Schoonheydt and co-workers reported O_2-activated Cu-ZSM-5 and Cu-MOR zeolites, which oxidize methane at 125 °C to yield methanol with 98% selectivity [12, 13]. The $[Cu_2(\mu\text{-}O)]^{2+}$ and $[Cu_3(\mu\text{-}O)_3]^{2+}$ species have been proposed as the active sites and debated for the past ten years [14–16], but no consensus has been reached as to its structure. The Co, Ni, Zn, and Rh sites in zeolites were also reported to be active for methane hydroxylation at low temperature [17–21].

The use of oxidants to form the active sites in metal-exchanged zeolites faces an insurmountable hurdle associated with the low availability of N_2O and the high temperature needed for O_2 activation. While spectroscopic observations have shown that the signature peaks of $[Cu_2(\mu\text{-}O)]^{2+}$ active sites are already seen at low temperature (100 and 175 °C for activation by N_2O and O_2, respectively) [14], it seems that low-temperature activations limit the methanol yield for unclear reasons (see Table 1). Despite this, mechanistic studies on the active site formation from oxidant activation are limited and strategies to circumvent the need of high-temperature activation are lacking. Thus, the formation of active sites, in addition to the methane's C–H bond cleavage, has become a new challenge in methane hydroxylation by metal-exchanged zeolites. In this chapter, I overview recent advances in the mechanistic

Table 1 Methane oxidation to methanol by Cu-zeolites at different activation temperatures

Catalysts[a]	Oxidant	CH_4 reaction temp (°C)	Methanol yield (μmol/g-cat) activation temperature (°C)				References
			200	350	450	≥550	
Cu-MOR (0.38)	O_2	200	–	15.9	25.8	37.5	[23]
Cu-MOR (0.21)	O_2	200	4.3	20.3	29.8	40.0	[23]
Cu-MOR (0.4)	O_2	150	–	21[b]	67	56[b]	[24]
Cu-MOR (0.4)	N_2O	150	–	37[b]	50[b]	97	[24]

[a] Values in the parentheses are Cu/Al ratios
[b] Approximated value from the referenced diagram

understanding of N_2O, O_2, and H_2O_2 activation on the reduced metal centers of Fe- and Cu-exchanged zeolites from a theoretical perspective. In addition, by employing density functional theory (DFT) calculations, I also present original findings on N_2O decomposition over two reduced Cu centers in the periodic structure of ZSM-5 zeolite.

2 Oxidative Activation of Fe-Exchanged Zeolites

2.1 *N_2O Decomposition on Fe^{II}-ZSM-5*

As the reduced metal center in Fe-exchanged zeolites has been recently found to be a mononuclear α-Fe^{II} species hosted on an Al pair within the 6-MR of ZSM-5 zeolite [6], here I discuss N_2O decomposition on the Fe^{II}-ZSM-5. As shown in Fig. 2a, the α-Fe^{II} center in ZSM-5 zeolite is coordinated to four lattice O atoms of the zeolite, forming a square planar geometry, while the $(Fe^{IV}{=}O)_\alpha$ active site has a pyramidal geometry with a Fe=O bond length of about 1.6 Å [6, 25].

N_2O decomposition into an N_2 molecule released to the atmosphere and an O ligand bound to the α-Fe^{II} center to form a $(Fe^{IV}{=}O)_\alpha$ active site is a straightforward process. Figure 2b shows detailed energy diagrams of N_2O decomposition on Fe^{II}-ZSM-5 reported recently by Yoshizawa and co-workers [26]. As shown in this figure, the α-Fe^{II} center (**IC**) prefers the quintet state. The N_2O molecule is adsorbed on the α-Fe^{II} center, forming a reactant complex (**RC**) with an adsorption energy of –7.3 kcal/mol and a Fe ··· O distance of 2.263 Å. In this reaction step, the geometry of N_2O remains the same as that in the gas phase, i.e. the N≡N and N–O bond lengths of N_2O are 1.126 and 1.186 Å, respectively. Then, the N–O bond is cleaved

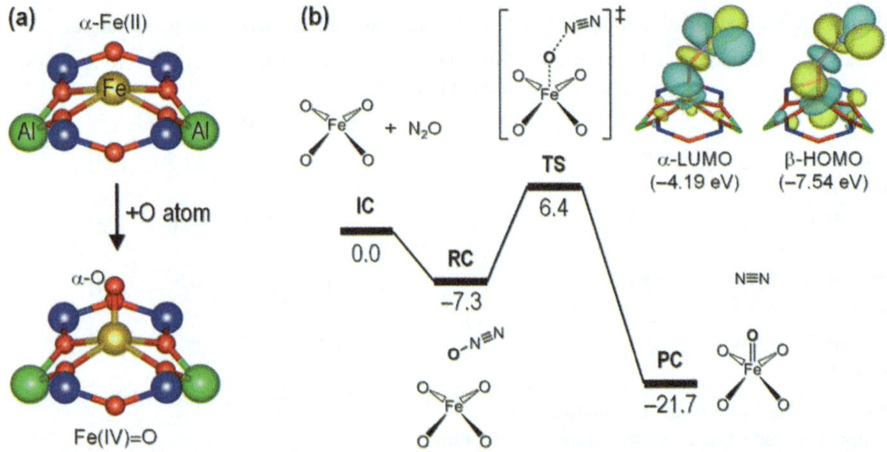

Fig. 2 **a** Reduced α-Fe^{II} and oxidized $(Fe^{IV}{=}O)_{\alpha}$ active sites in the six-membered ring (6-MR) of ZSM-5 zeolite. **b** Reaction energy diagrams for N_2O decomposition on the periodic structure of Fe^{II}-ZSM-5 in the quintet ground state. Energies are in kcal/mol. **IC**, **RC**, **TS**, and **PC** stand for initial complex, reactant complex, transition state, and product complex, respectively. Reproduced from Ref. [26] by permission of The Royal Society of Chemistry

via a transition state (**TS**) with a separated N ··· O distance of 1.468 Å, a bent N–N–O angle of 142°, an approaching Fe ··· O distance of 1.819 Å, and an activation energy of 13.7 kcal/mol, the latter of which agrees very well with the previous DFT results without dispersion correction (15.1 or 16.5 kcal/mol) [25, 27] and with the experimental value (14.1 kcal/mol) [28]. At the transition state, the Bader charge and spin density of N_2O are −0.3*e* and −0.21, respectively, indicating a one-electron transfer from the Fe^{II} center to a low-lying π* orbital of N_2O. Figure 2b shows the lowest unoccupied molecular orbital (LUMO) for the α electron and the highest occupied molecular orbital (HOMO) for β electron of **TS** calculated at the B3LYP level of theory (see Ref. [26] for the cluster model and computational methods used). As can be seen from these two MOs, the O–p_z orbital of N_2O makes antibonding (LUMO) and bonding (HOMO) interactions with the Fe–d_{z^2} orbital, which would, respectively, result in the 3σ* and 2σ orbitals of $Fe^{IV}{=}O$ [25]. In the last step, another one-electron transfer completes the two-electron oxidation process, forming an N_2 molecule and a $(Fe^{IV}{=}O)_{\alpha}$ active site in the quintet state.

2.2 H_2O_2 Decomposition on $[Fe^{III}–(\mu O)_2–Fe^{III}]$-ZSM-5

Following the work of Hutchings and co-workers [8], who proposed a dinuclear $[(H_2O)(HO)–Fe–(\mu OH)_2–Fe–(OH)(H_2O)]^{2+}$ as the active site in the liquid-phase methane hydroxylation by Fe-ZSM-5, Pidko and co-workers recently suggested a reduced Fe center model $[(H_2O)_2–Fe^{III}–(\mu O)_2–Fe^{III}–(H_2O)_2]^{2+}$ (see Fig. 3) which is

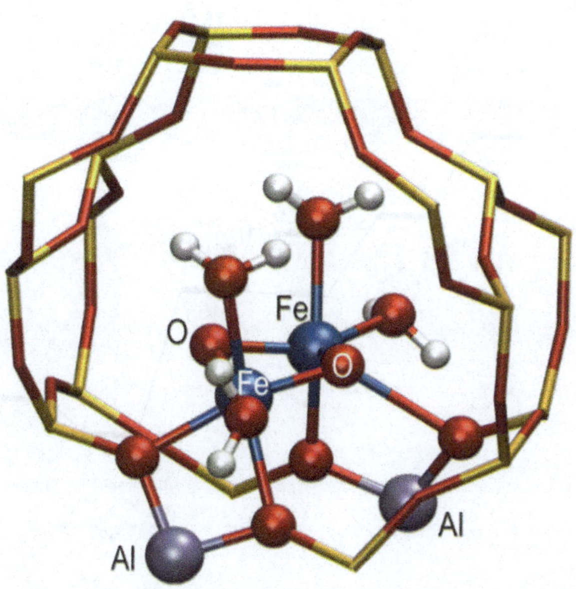

Fig. 3 A reduced Fe center model $[(H_2O)_2–Fe^{III}–(\mu O)_2–Fe^{III}–(H_2O)_2]^{2+}$ on the eight-membered ring (8-MR) of ZSM-5 zeolite. Adapted with permission from Ref. [29]. Further permissions related to this figure should be directed to American Chemical Society

plausible for the active site formation using H_2O_2 [29]. The authors demonstrated that this Fe center may lead to a number of possible active site structures after the decomposition of H_2O_2, namely (**1**) $[(H_2O)–Fe^{III}(HOOH)–(\mu O)_2–Fe^{III}(HOH)–(H_2O), H_2O]^{2+}$, (**2**) $[(H_2O)–Fe^{III}(OOH)–(\mu O)–Fe^{III}(OH)–(H_2O)]^{2+}$, and (**3**) $[(H_2O)_2–Fe^{IV}–(\mu O)_2–Fe^{IV}O–(H_2O)]^{2+}$ (Scheme 1), which catalyze the Fenton-type, heterolytic, and homolytic activation of methane, respectively. The bond dissociation energy (BDE) of H_2O_2 is only 34 kcal/mol. Thus, cleaving the HO–OH bond is expected to be slightly easier than breaking the N_2–O bond (BDE = 42 kcal/mol).

Figure 4 (dotted line) shows detailed energy diagrams for the formation of **1**, where H_2O_2 is simply adsorbed on the reduced Fe centers and then bound to one of the Fe centers replacing a H_2O ligand which is now uncoordinated. In the formation of **2** (Fig. 4, black solid lines), the coordinated H_2O_2 is deprotonated by the bridging O site to form a bridging OH group and a terminal OOH ligand as a reaction intermediate.

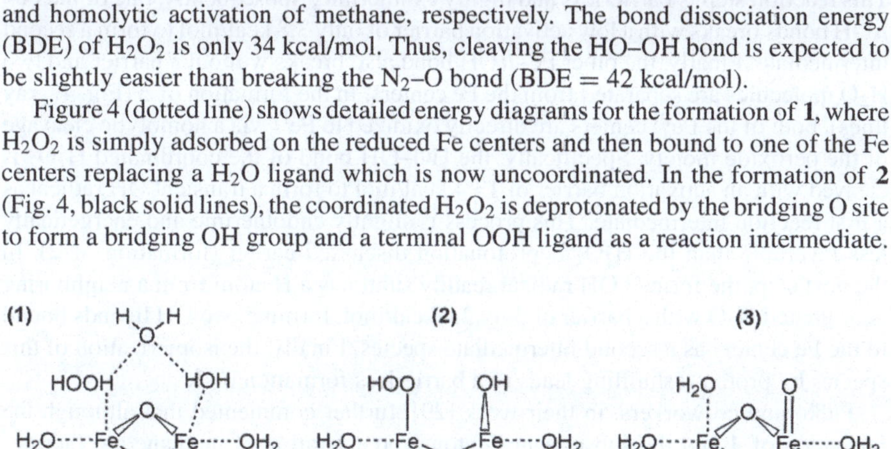

Scheme 1 Three possible active site structures of Fe-ZSM-5 for methane activation through the (**1**) Fenton-type, (**2**) heterolytic, and (**3**) homolytic mechanisms. Reproduced with permission from Ref. [29]. Copyright 2018 American Chemical Society

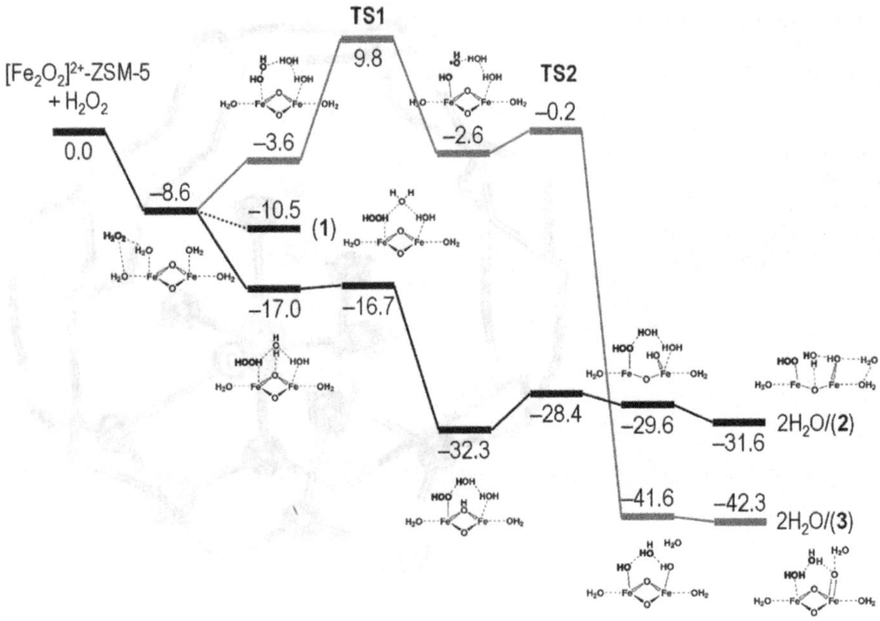

Fig. 4 Reaction energy diagrams of H_2O_2 decomposition on $[Fe^{III}–(\mu O)_2–Fe^{III}]^{2+}$-ZSM-5 zeolite in the corresponding ground state. Energies are in kcal/mol. Reproduced with permission from Ref. [29]. Further permissions related to this figure should be directed to American Chemical Society

This reaction step is barrierless and highly exothermic. Subsequently, one of the Fe–μOH bonds breaks with a low activation barrier of only 3.8 kcal/mol to form a second intermediate. Finally, the other Fe–μOH bond also breaks without a barrier and two H_2O molecules are separated from the Fe centers. In the formation of **3** (Fig. 4, gray lines), both of the Fe^{III} centers are directly oxidized to Fe^{IV} via a homolytic cleavage of the peroxide moiety. Specifically, the HO–OH bond of the coordinated H_2O_2 is cleaved with an activation barrier of 13.4 kcal/mol to form a transient OH radical as a first reaction intermediate. This process is slightly endothermic and energetically less favorable than the H_2O_2 deprotonation discussed earlier (formation of **2**). In the next step, the formed OH radical readily subtracts a H atom from a neighboring coordinated H_2O with a barrier of only 2.4 kcal/mol, forming two OH ligands bound to the Fe centers as a second intermediate species. Finally, the isomerization of this species by proton reshuffling leads to a barrierless formation of **3**.

Pidko and co-workers, in their work [29], further commented that although the formation of **1** and the subsequent Fenton-type activation of methane are thermodynamically the most favorable, Fenton-type reaction is in general undesired, due to its difficulty to control and its tendency to decrease the selectivity of the overall process. On the other hand, the heterolytic and homolytic activation of methane by **2** and **3**, respectively, are energetically reasonable, suggesting that **2** and **3** are more plausible as the active site of Fe-ZSM-5 than **1**.

3 Oxidative Activation of Cu-Exchanged Zeolites

3.1 N_2O Decomposition on $2Cu^I$-ZSM-5

Following the previous work by Woertink et al. suggesting that the $[Cu_2(\mu\text{-}O)]^{2+}$ formation from N_2O decomposition is already observed at 100 °C [14], Tsai et al. carried out kinetics experiments to determine the activation energy of N_2O activation on Cu-ZSM-5 by measuring the reaction rate at six different temperatures ranging from 25 to 100 °C [30]. The measured activation energy from the Arrhenius plot was reported to be $E_a = 2.5$ kcal/mol, which agree very well with the DFT-calculated apparent activation barrier (2.0 kcal/mol) [30].

Figure 5 shows detailed energy diagrams of N_2O decomposition on a small cluster model of $2Cu^I$-ZSM-5 reported by Tsai et al., who used the B3LYP functional without van der Waals (vdW) dispersive correction for the calculations [30]. The authors suggested three modes of N_2O binding on the two Cu^I centers: μ-1,1-O, μ-1,3-O,N, and η^1-N binding modes in the closed-shell singlet (CSS) ground state. Although the η^1-N binding mode results in the highest binding energy, it lacks a proper reaction coordinate to form a bridging oxo on the Cu centers. Also, the μ-1,3-O,N binding mode, which results in the second highest binding energy, requires a higher apparent O–N_2 activation energy (5 kcal/mol) than that via the μ-1,1-O binding mode (2 kcal/mol). Thus, the μ-1,1-O binding mode was chosen as the reactant complex (**RC**) of the reaction. In **RC**, Tsai et al. reported that the O atom of N_2O is coordinated to the two Cu^I centers with different Cu–O bond lengths (2.0 and 2.5 Å, see Fig. 5). However, our DFT calculations using the periodic structure of ZSM-5 zeolite and employing the PBE functional and the vdW-D2 method of dispersion correction (Fig. 6) show that the Cu–O bonds are similar in length (about 2.0 Å) and the Cu···Cu distance is shorter (2.55 Å). Moreover, our computational results show that N_2O is bound with a binding energy of −27.3 kcal/mol, which is

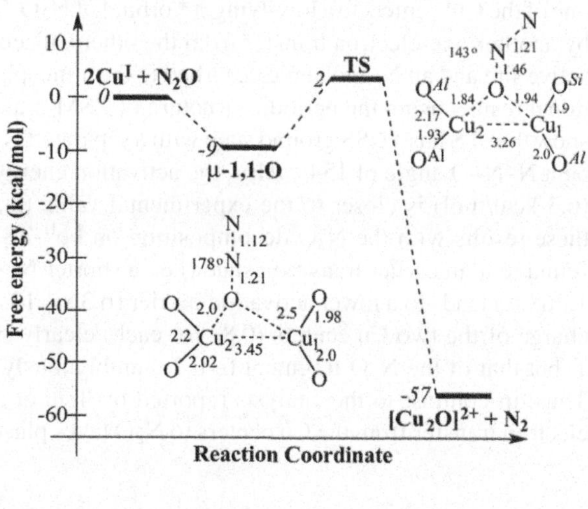

Fig. 5 Reaction energy diagrams of N_2O decomposition on a small cluster model of $2Cu^I$-ZSM-5 zeolite in the corresponding ground state. Energies are in kcal/mol. Adapted with permission from Ref. [30]. Copyright 2014 American Chemical Society

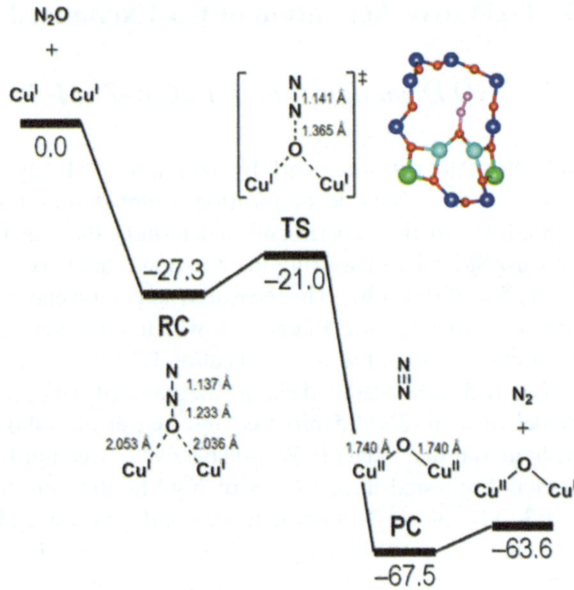

Fig. 6 Reaction energy diagrams of N_2O decomposition on the periodic structure of $2Cu^I$-ZSM-5 zeolite (T3/T3 Al pair site) in the corresponding ground state (initial, **RC**, **TS**: closed-shell singlet state; **PC** and final: triplet state). Energies are in kcal/mol. Computational methods used for calculating the diagram are explained elsewhere [26]

much stronger than that reported by Tsai et al. (−9 kcal/mol). This is possibly due to the shorter Cu–O bonds and the vdW effects from the zeolite framework. After the binding, the N–O bond of N_2O is slightly elongated from 1.186 to 1.233 Å, while the N–N–O angle remains unchanged (i.e. 178°).

The N–O bond is then cleaved via a **TS** structure in the open-shell singlet (OSS) state with a separated N ··· O distance of 1.46 Å and a bent N–N–O angle of 143° (Fig. 5) [30]. The activation energy measured from **RC** is 11 kcal/mol, while the apparent activation energy measured from the gas-phase state is 2 kcal/mol. According to Tsai et al. [30], the change in Mulliken charges on N_2O (**RC** 0.04, **TS** −0.40, **PC** −0.93) indicates that during the cleavage, one electron is transferred from one of the Cu^I centers to a low-lying π* orbital of N_2O. The reaction is then completed by another one-electron transfer from the other Cu center, forming a $[Cu^{II}{}_2(\mu\text{-}O)]^{2+}$ active site and an N_2 molecule desorbed to the atmosphere. In contrast, our computational results using the periodic structure of ZSM-5 and the PBE functional (Fig. 6) show that **TS** has a CSS ground state with a separated N ··· O distance of 1.365 Å and a bent N–N–O angle of 154°. Also, the activation energy for the N_2–O bond cleavage (6.3 kcal/mol) is closer to the experimental value (2.5 kcal/mol) [30]. Comparing these results with the N_2O decomposition on Fe^{II}-ZSM-5 discussed in Sect. 2.1, I found that an earlier transition state (i.e., a shorter N ··· O distance of **TS**: 1.365 vs. 1.468 Å) leads to a lower activation barrier (6.3 vs. 13.7 kcal/mol). At **TS**, the Bader charge of the two Cu centers (0.8*e* for each) clearly shows an oxidation state of + 1, but that of the N_2O fragment (−0.2*e*) ambiguously indicate an electron transfer. Thus, in contrast to the analysis reported by Tsai et al. [30], I suggest that a two-electron transfer from the Cu centers to N_2O takes place at the product complex (**PC**, triplet ground state), where the N–O bond is completely cleaved and an N_2 molecule is formed and readily desorbed with a desorption energy of only 3.9 kcal/mol.

3.2 O_2 Activation on $2[Cu^I{}_2]$-MOR and $[Cu^{III}{}_2Cu^I(MO)]$-MOR

In contrast to the N_2O decomposition, O_2 activation requires four electrons to break the O=O bond. Unfortunately, the reduced $Cu^I \cdots Cu^I$ centers in Cu-zeolites can provide only two electrons. Thus, two additional electrons from two spectator Cu^I sites are required. Detailed mechanism of this reaction was reported by Smeets et al. [31], who experimentally found a μ-η^2:η^2-peroxo-$Cu^{II}{}_2$ species as a precursor for the formation of mono(μ-O)$Cu^{II}{}_2$ active species. The main issue associated with this mechanism, however, is the fate of the second O atom after the peroxo bond of the precursor is cleaved. Smeets et al. suggested that the second O atom, instead of forming a second $[Cu_2(\mu\text{-}O)]^{2+}$ active species, interacts with the zeolite lattice that acts as a reservoir for mobile O atoms [31]. However, a recent DFT study by the Yoshizawa's group showed that when the second O atom interacts with the lattice O atom and forms a Si–O–O–Si fragment (**PC'** in Fig. 7, gray lines), the activation energy required for breaking the μ-η^2:η^2-peroxo bond of the precursor is extremely high (59.7 kcal/mol) [32]. On the other hand, the simultaneous formation of two

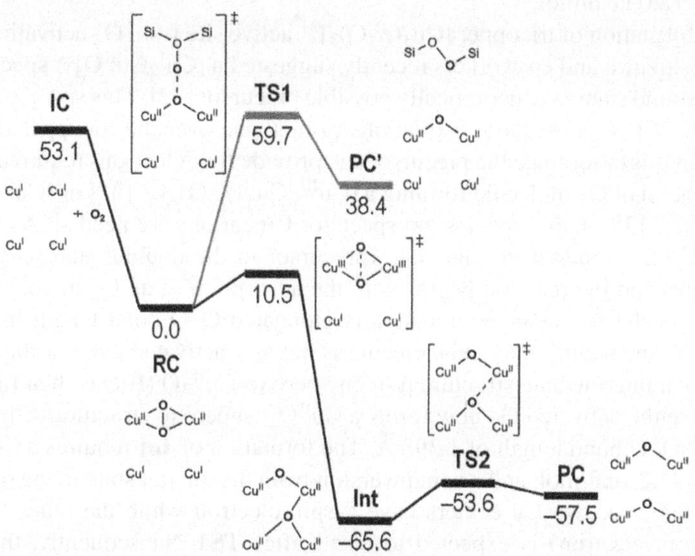

Fig. 7 Reaction energy diagrams for two distinct reaction pathways of O_2 activation on $2[Cu^I{}_2]^{2+}$-MOR in the corresponding ground state. Energies are in kcal/mol. Reproduced with permission from Ref. [32]. Copyright 2018 American Chemical Society

neighboring $[Cu_2(\mu\text{-}O)]^{2+}$ active sites via a transition structure of $[Cu^{III}\text{–}(\mu O)_2\text{–}Cu^{III}]^{2+}$ is energetically more favorable with an O–O bond activation barrier of only 10.5 kcal/mol [32].

Figure 7 (black lines) shows detailed energy diagrams of O_2 activation on $2[Cu^I{}_2]$-MOR, where two $[Cu_2(\mu\text{-}O)]^{2+}$ active sites are simultaneously formed. In this mechanism, there are two $Cu^I \cdots Cu^I$ pairs hosted on four Al tetrahedra within the 8-MR side-pocket windows of MOR. These windows are neighboring to each other and separated by about 4 Å. The initial complex (**IC**) is formed in the triplet ground state since the Cu^I centers have a CSS ground state and O_2 in the gas phase (O=O = 1.207 Å) has a triplet ground state. Subsequently, the formation of $\mu\text{-}\eta^2\text{:}\eta^2$-peroxo-$Cu^{II}{}_2$ precursor (O–O = 1.465 Å) in the OSS ground state is extremely exothermic with a binding energy of −53.1 kcal/mol. Here, one of the $Cu^I \cdots Cu^I$ pairs is oxidized to $Cu^{II} \cdots Cu^{II}$ while the other one remains unchanged. The peroxo bond is then cleaved via a first transition state (**TS1**) having a structure of bis(μ-O)$Cu^{III}{}_2$ (O ⋯ O = 2.201 Å) in the CSS state. The activation energy required for this process is 10.5 kcal/mol, which is expectedly higher than that for the N_2–O cleavage discussed in Sect. 3.1. At this step, a four-electron process of O=O bond cleavage is already complete. However, to form two $[Cu_2(\mu\text{-}O)]^{2+}$ active sites, one of the separated O atoms should migrate to the neighboring 8-MR side-pocket window through the formation of an intermediate species (**Int**). This intermediate structure is energetically very stable in the triplet ground state with all Cu centers having an oxidation state of +2. Finally, the formation of two $[Cu^{II}{}_2(\mu\text{-}O)]^{2+}$ active sites in the triplet state then takes place via a second transition state (**TS2**), requiring an activation energy of 12.0 kcal/mol.

In the formation of tricopper $[Cu_3(\mu\text{-}O)_3]^{2+}$ active site from O_2 activation on Cu-MOR, Yoshizawa and co-workers recently suggested a $[Cu^I{}_2Cu^{II}O]^{2+}$ species in the doublet ground state as a theoretically possible precursor [32]. This species is formed as a result of two consecutive oxidations of methane over the tricopper active site [33, 34]. In this case, since the precursor can provide four electrons required to break the O=O bond of O_2 molecule forming a $[Cu^{III}{}_2Cu^{II}(\mu\text{-}O)_3]^{2+}$ [32] or $[Cu^{II}{}_2Cu^{III}(\mu\text{-}O)_2(\mu\text{-}O^{\cdot})]^{2+}$ [35] active species, no spectator Cu cations are needed. As shown in Fig. 8, **IC** has a quartet ground state (precursor in the doublet state + O_2 in the triplet state) and the reaction begins with the adsorption of an O_2 molecule on one Cu center of the precursor with a slightly elongated O–O bond length from 1.207 to 1.260 Å, suggesting that the molecule is already activated even at this stage of reaction. An intermediate structure $\mu\text{-}\eta^2\text{:}\eta^2$-peroxo-$Cu^{II}{}_3O$ (**Int**) is then formed via **TS1**, where the activated O_2 now forms a $Cu^{II}O_2{}^{\cdot}$ superoxo structure with a further elongated O–O bond length of 1.398 Å. The formation of **Int** requires an activation energy of 24.2 kcal/mol, and a spin inversion from the quartet state to the open-shell doublet state (i.e. two Cu centers have α-spin electron while the other Cu center has a β-spin electron) is expected to occur after **TS1**. Subsequently, the peroxo bond of **Int** is cleaved via **TS2** with a cleaved O ⋯ O distance of 1.810 Å to form the $[Cu_3(\mu\text{-}O)_3]^{2+}$ active site in the quartet ground state. Given another activation barrier of 10.4 kcal/mol for this cleavage, we obtain an effective activation barrier of 31.6 kcal/mol.

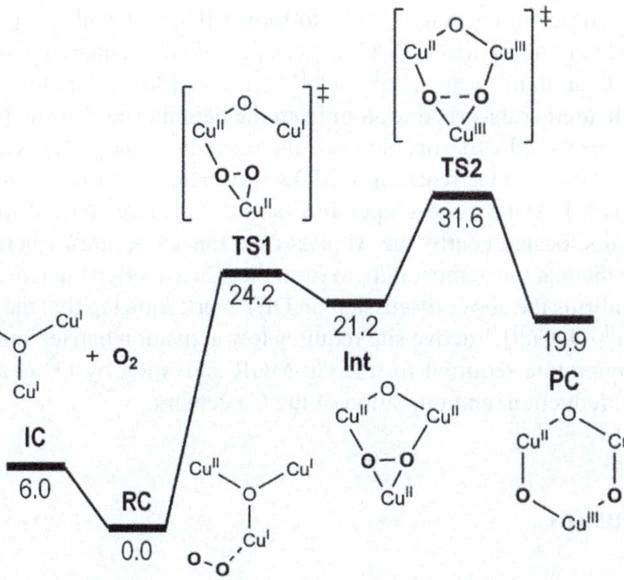

Fig. 8 Reaction energy diagrams of O_2 activation on $[Cu^I{}_2Cu^{II}(\mu\text{-}O)]^{2+}$-MOR in the corresponding ground state. Energies are in kcal/mol. Reproduced with permission from Ref. [32]. Copyright 2018 American Chemical Society

Recently, Ikuno et al. systematically described that the O=O bond cleavage of O_2 forming the $[Cu_3(\mu\text{-}O)_3]^{2+}$ active site actually occurs at low temperature (50–200 °C), but dehydration, auto-reduction, and subsequent migration of the Cu centers at high temperature are required beforehand [36]. As shown in Fig. 9, at the initial stage, two $[Cu^{II}OH]^+$ species derived from a dehydration of $[Cu^{II}(H_2O)_n]^{2+}$ are condensed

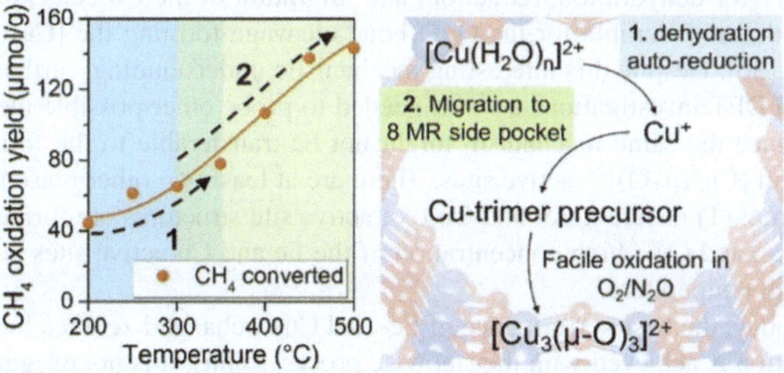

Fig. 9 Elementary steps in the formation of $[Cu_3(\mu\text{-}O)_3]^{2+}$ active site in Cu-MOR: (1) Moderate-temperature dehydration and auto-reduction of the Cu centers, (2) high-temperature migration of the reduced Cu centers to form a favorable precursor, and (3) low-temperature O_2 or N_2O activation forming the $[Cu_3(\mu\text{-}O)_3]^{2+}$ active site. Adapted with permission from Ref. [36]. Copyright 2019 American Chemical Society

at moderate temperature (up to 300 °C) to form a $[Cu^{II}–O–Cu^{II}]^{2+}$ species. Raising the temperature to more than 450 °C reduces the two Cu centers to Cu^{I} and desorbs the bridging O atom to form a $[Cu^{I} \cdots Cu^{I}]^{2+}$ species [36] and an O_2 molecule after rebound with another desorbed O atom from the neighboring Cu site [37]. The high temperature seems to be reasonable since the formation energy for $[Cu^{II}–O–Cu^{II}]^{2+} \rightarrow [Cu^{I} \cdots Cu^{I}]^{2+} + ½ O_2$ reaction in MOR was calculated to be quite high (about 55 kcal/mol) [33]. At the same temperature, some of the reduced Cu^{I} species migrate to the Cu^{I} pairs located nearby the Al pairs (i.e., the active site), where O_2 and $^{\cdot}OH$ then oxidize them at low temperature to form the $[Cu_3(\mu\text{-}O)_3]^{2+}$ active site [36]. This therefore confirms the above discussion on DFT work showing that the O_2 activation forming $[Cu^{II}–O–Cu^{II}]^{2+}$ active site requires low activation barrier and clarifies that the high temperature required for the Cu-MOR activation by O_2 is mainly due to dehydration, reduction, and migration of the Cu centers.

4 Conclusion

In this chapter, I have overviewed recent topics in the mechanistic understanding of oxidative activation of Fe- and Cu-exchanged zeolites from a theoretical point of view. While the N_2O and H_2O_2 decompositions are straightforward two-electron cleavage processes, the O_2 activation is a four-electron cleavage process involving more complicated mechanisms. Despite such a difference, we have shown that the activation energies required for N_2O, H_2O_2, and O_2 activations on the reduced Fe and Cu centers discussed here (except on the $[Cu_3O]^{2+}$) are lower than 20 kcal/mol, implying that the activation can in principle be achieved at low temperatures. This is actually consistent with the recent experimental work on XAS and UV–vis spectroscopies [36] although it is known that a high-temperature activation of Cu-MOR by O_2 is more favored for higher yield of methanol. Such a thermal pre-treatment is necessary for dehydration, reduction, and migration of the Cu centers to form a precursor that is favorable for the O=O bond cleavage forming the $[Cu_3(\mu\text{-}O)_3]^{2+}$ active site [36]. Despite this interesting mechanistic understanding, further spectroscopic and DFT investigations are still needed to probe other possible mechanistic reasons, since the same mechanism might not be transferable to the formation of $[FeO]^{2+}$ and $[Cu_2(\mu\text{-}O)]^{2+}$ active sites. There are at least two other reasons that are possibly true: (1) different Fe- and Cu-oxo active site structures are formed at high temperature, and (2) a high concentration of the Fe and Cu active sites is achieved only at high temperature.

Until today, the best performance of Fe- and Cu-exchanged zeolites for methane hydroxylation is achieved with the stepwise process. Thus, it is not exaggerating to say that this process will still be the main topic of discussion in methane hydroxylation by metal-exchanged zeolite catalysts. However, the oxidative activation step involved in this process will be an endless bottleneck if massive investigations are not directed to address this challenge. Beside the O_2 activation, the H_2O decomposition to form the Cu active sites [38] is also highly desirable, as water is abundant and the reaction produces H_2, too. It is therefore important to spectroscopically and computationally study the mechanism of H_2O decomposition on Cu-zeolites.

Acknowledgements The computations were performed at the Research Institute for Information Technology (Kyushu University).

References

1. Narsimhan K, Iyoki K, Dinh K, Román-Leshkov Y (2016) ACS Cent Sci 2:424–429
2. Dinh KT, Sullivan MM, Narsimhan K, Serna P, Meyer RJ, Dincă M, Román-Leshkov Y (2019) J Am Chem Soc 141:11641–11650
3. Panov GI, Sobolev VI, Kharitonov AS (1990) J Mol Catal 61:85–97
4. Sobolev VI, Dubkov KA, Panna OV, Panov GI (1995) Catal Today 24:251–252
5. Panov GI, Sobolev VI, Dubkov KA, Parmon VN, Ovanesyan NS, Shilov AE, Shteinman AA (1997) React Kinet Catal Lett 61:251–258
6. Snyder BER, Vanelderen P, Bols ML, Hallaert SD, Böttger LH, Ungur L, Pierloot K, Schoonheydt RA, Sels BF, Solomon EI (2016) Nature 536:317–321
7. Bols ML, Hallaert SD, Snyder BER, Devos J, Plessers D, Rhoda HM, Dusselier M, Schoonheydt RA, Pierloot K, Solomon EI, Sels BF (2018) J Am Chem Soc 140:12021–12032
8. Hammond C, Forde MM, Ab Rahim MH, Thetford A, He Q, Jenkins RL, Dimitratos N, Lopez-Sanchez JA, Dummer NF, Murphy DM, Carley AF, Taylor SH, Willock DJ, Stangland EE, Kang J, Hagen H, Kiely CJ, Hutchings GJ (2012) Angew Chemie Int Ed 51:5129–5133
9. Hammond C, Dimitratos N, Lopez-Sanchez JA, Jenkins RL, Whiting G, Kondrat SA, ab Rahim MH, Forde MM, Thetford A, Hagen H, Stangland EE, Moulijn JM, Taylor SH, Willock DJ, Hutchings GJ (2013) ACS Catal 3:1835–1844
10. Hammond C, Hermans I, Dimitratos N (2015) ChemCatChem 7:434–440
11. Xiao P, Wang Y, Nishitoba T, Kondo JN, Yokoi T (2019) Chem Commun 55:2896–2899
12. Groothaert MH, Smeets PJ, Sels BF, Jacobs PA, Schoonheydt RA (2005) J Am Chem Soc 127:1394–1395
13. Smeets PJ, Groothaert MH, Schoonheydt RA (2005) Catal Today 110:303–309
14. Woertink JS, Smeets PJ, Groothaert MH, Vance MA, Sels BF, Schoonheydt RA, Solomon EI (2009) Proc Natl Acad Sci U S A 106:18908–18913
15. Vanelderen P, Snyder BER, Tsai M-L, Hadt RG, Vancauwenbergh J, Coussens O, Schoonheydt RA, Sels BF, Solomon EI (2015) J Am Chem Soc 137:6383–6392
16. Grundner S, Markovits MAC, Li G, Tromp M, Pidko EA, Hensen EJM, Jentys A, Sanchez-Sanchez M, Lercher JA (2015) Nat Commun 6:7546
17. Beznis N V., Weckhuysen BM, Bitter JH (2010) Catal Letters 136:52–56
18. Shan J, Huang W, Nguyen L, Yu Y, Zhang S, Li Y, Frenkel AI, Tao F (2014) Langmuir 30:8558–8569
19. Mahyuddin MH, Yoshizawa K (2018) Catal Sci Technol 8:5875–5885
20. Oda A, Ohkubo T, Yumura T, Kobayashi H, Kuroda Y (2019) Inorg Chem 58:327–338
21. Shan J, Li M, Allard LF, Lee S, Flytzani-Stephanopoulos M (2017) Nature 551:605–608
22. Tomkins P, Ranocchiari M, van Bokhoven JA (2017) Acc Chem Res 50:418–425
23. Park MB, Ahn SH, Ranocchiari M, van Bokhoven J (2017) ChemCatChem 9:3705–3713
24. Kim Y, Kim TY, Lee H, Yi J (2017) Chem Commun 53:4116–4119
25. Mahyuddin MH, Shiota Y, Staykov A, Yoshizawa K (2017) Inorg Chem 56:10370–10380
26. Mahyuddin MH, Shiota Y, Yoshizawa K (2019) Catal Sci Technol 9:1744–1768
27. Li G, Pidko EA, Filot IAW, van Santen RA, Li C, Hensen EJM (2013) J Catal 308:386–397
28. Kondratenko E V., Pérez-Ramírez J (2006) J Phys Chem B 110:22586–22595
29. Szécsényi Á, Li G, Gascon J, Pidko EA (2018) ACS Catal 8:7961–7972

30. Tsai M-L, Hadt RG, Vanelderen P, Sels BF, Schoonheydt RA, Solomon EI (2014) J Am Chem Soc 136:3522–3529
31. Smeets PJ, Hadt RG, Woertink JS, Vanelderen P, Schoonheydt RA, Sels BF, Solomon EI (2010) J Am Chem Soc 132:14736–14738
32. Mahyuddin MH, Tanaka T, Staykov A, Shiota Y, Yoshizawa K (2018) Inorg Chem 57:10146–10152
33. Mahyuddin MH, Tanaka T, Shiota Y, Staykov A, Yoshizawa K (2018) ACS Catal 8:1500–1509
34. Zheng J, Lee I, Khramenkova E, Wang M, Peng B, Gutierrez O, Fulton JL, Camaioni D, Khare R, Jentys A, Haller G, Pidko E, Sanchez-Sanchez M, Lercher J (2020) Chem – A Eur J 26:7563–7567
35. Mahyuddin MH, Staykov A, Saputro AG, Agusta MK, Dipojono HK, Yoshizawa K (2020) J Phys Chem C 124:18112–18125
36. Ikuno T, Grundner S, Jentys A, Li G, Pidko E, Fulton J, Sanchez-Sanchez M, Lercher JA (2019) J Phys Chem C 123:8759–8769
37. Sushkevich VL, Smirnov A V., van Bokhoven JA (2019) J Phys Chem C 123:9926–9934
38. Sushkevich VL, Palagin D, Ranocchiari M, van Bokhoven JA (2017) Science 356:523–527

Dynamics and Energetics of Methane on the Surfaces of Transition Metal Oxides

Yuta Tsuji, Masashi Saito, and Kazunari Yoshizawa

Abstract In the very early part of this chapter, we review the Langmuir model of adsorption to understand the dynamics of the competition between adsorption and desorption of methane on a surface. Then, the kinetics model introduced is further developed so that one can apply it to the dissociation dynamics of the C–H bond of methane on surface whether it happens in a high-pressure limit or low-pressure one. In the latter limit, an interesting concept of the apparent activation energy is introduced and its correlation with the desorption barrier or the adsorption energy is clarified. In order to reduce the apparent activation energy associated with the dissociation of the C–H bond of methane on a surface, one needs to find a surface which enjoys a substantial affinity to methane; such a candidate can be found in a class of metal oxides. In such a situation, what is so-called trapping-mediated mechanism can apply. The (101) surface of PdO and the (110) and (100) surfaces of IrO_2 and RuO_2 have been experimentally and theoretically characterized as a good candidate for such a reaction mechanism. We carry out a comprehensive survey of their surface features, stressing an importance of possessing a coordinatively unsaturated (cus) metal site as well as a cus O site nearby on the surface so that a H atom of methane can be abstracted by the cus O site and the resultant CH_3 fragment can be significantly stabilized by an interaction with the cus metal site. Taking the IrO_2 (110) surface as an example, we conduct a thorough scrutiny on the electronic structure of methane on the surface at the level of the extended Hückel method, which can provide us with a qualitatively correct insight into orbital interaction. We make good use of a band decomposition technique called crystal orbital overlap population (COOP) to unveil some important orbital interactions between the C–H bond dissociated and the surface cus Ir and O atoms.

Keywords Adsorption · Orbital interaction · Crystal orbital overlap population · Metal oxide · Dative bond

Y. Tsuji · M. Saito · K. Yoshizawa (✉)
Institute for Materials Chemistry and Engineering and IRCCS, Kyushu University, Nishi-ku, Fukuoka 819-0395, Japan
e-mail: kazunari@ms.ifoc.kyushu-u.ac.jp

K. Yoshizawa (ed.), *Direct Hydroxylation of Methane*,
https://doi.org/10.1007/978-981-15-6986-9_6

1 Introduction

In recent years, unconventional natural-gas resources, such as methane hydrate [1] and shale gas [2], have attracted much attention. Along with the progress of the drilling technology, methane output is expected to increase significantly in the future. Therefore, the development of effective utilization methods of methane as a chemical raw material to transform it to value-added products is of significant importance [3]. To this end, one needs to activate the C–H bond of methane as the first step; however, such an activation scheme is challenging because methane's C–H bonds are robust, as exemplified by a large bond dissociation energy of 104 kcal/mol. Nevertheless, human beings have achieved a variety of ways to activate methane, be it homogeneous [4] or heterogeneous catalysts [5], as one could see in other chapters of this book. Among them is the metal oxide catalyst. Detailed insights into methane activation on oxide surfaces are expected to allow us to design a new catalyst or improve existing catalytic processes. In this chapter, for starters, we will make a comprehensive survey of methane activation on various well-defined metal oxide surfaces and later we will take a closer look at how actually methane is activated on the surface, taking the surface of a certain metal oxide, IrO_2, as an example.

2 Kinetics of Methane on Surface

2.1 Langmuir Model

It would be instructive to begin with a basic surface chemistry so that the reader could better understand the dynamics of the adsorption of methane on the surface. We will begin with a simple yet effective model named after Irving Langmuir [6], wherein many things are assumed; for example, the surface consists of a series of identical sites, there is no interaction between the adsorbate molecules whether they are on the surface or in the gas phase, a randomly moving adsorbate molecule plunges to the surface, the adsorbate is trapped by the surface with a certain probability, the adsorbate does not affect the adsorption abilities of the other sites, and the adsorbate cannot be adsorbed at the already occupied site but will be scattered. These assumptions seem to be too simplified or extreme; however, they are mostly fulfilled in the case of the adsorption of methane on a clean surface [7].

Figure 1 shows schematic representations of the Langmuir model describing the adsorption and desorption dynamics of a molecule, say, methane, represented by a shaded ball. When the total area of the surface is S, that of the occupied sites S_{ad}, and that of the unoccupied sites S_0, or $S - S_{ad}$, the desorption rate can be equal to $k_{des} S_{ad}$, where k_{des} is the rate constant for desorption. As for the adsorption rate, since it depends on the pressure of the adsorbate, P, as well as the area of the unoccupied sites, it reads $k_{ad} S_0 P$, where k_{ad} is the adsorption rate constant. Since $S_0 = S - S_{ad}$,

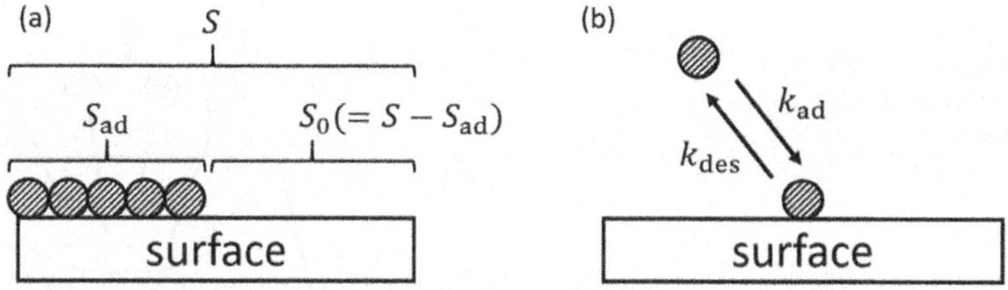

Fig. 1 Schematic representation of the Langmuir model. In **a** the area of the occupied sites is denoted as S_{ad}, while that of the unoccupied sites S_0. In **b** the dynamics of the adsorption and desorption is shown, where the rate constants for those are denoted as k_{ad} and k_{des}, respectively

it follows that the adsorption rate is written as $k_{ad}P(S - S_{ad})$. It would be reasonable to assume that there is equilibrium between adsorption and desorption, resulting in.

$$k_{des}S_{ad} = k_{ad}P(S - S_{ad}). \tag{1}$$

Note that the coverage of the adsorbate is defined as $\theta = S_{ad}/S$. Using θ, we can obtain

$$k_{des}\theta = k_{ad}P(1 - \theta). \tag{2}$$

After some algebra, it would not be that difficult to arrive at

$$\theta = \frac{k_{ad}P}{k_{des} + k_{ad}P}. \tag{3}$$

This is what is called the Langmuir adsorption isotherm.

Let us take a closer look at the rate constants for adsorption and desorption. When one wants to consider the adsorption process, so-called Hertz-Knudsen equation [8, 9] is helpful, in which the flux of N molecules impinging on a surface area, A, is modeled (see Fig. 2). The collision frequency per unit area can be written as [10, 11].

$$Z_{coll} = \frac{1}{A}\frac{dN}{dt} = \frac{P}{\sqrt{2\pi mkT_g}}, \tag{4}$$

where m is the mass of the adsorbate molecule, k the Boltzmann constant, and T_g the temperature of gas. Once Z_{coll} is multiplied by $(1 - \theta)$, one finds the resultant formula is none other than the adsorption rate, leading to

$$\frac{P}{\sqrt{2\pi mkT_g}}(1 - \theta) = k_{ad}P(1 - \theta). \tag{5}$$

This is why we have

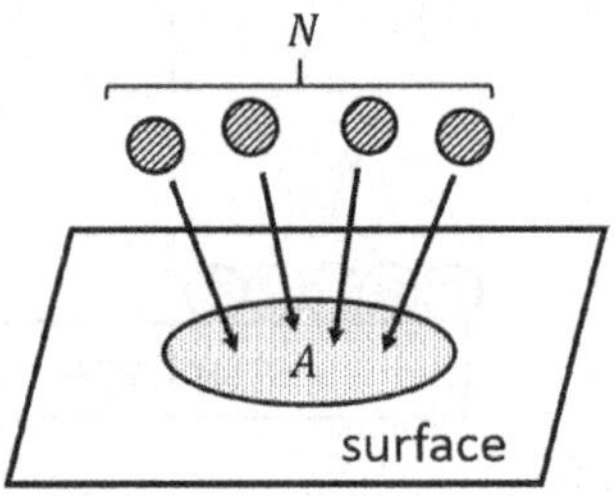

Fig. 2 Schematic representation of the flux of N molecules impinging on a surface area, A

$$k_{\mathrm{ad}} = \frac{1}{\sqrt{2\pi mkT_g}}. \tag{6}$$

Next, we turn to the desorption rate, $k_{\mathrm{des}}S_{\mathrm{ad}}$. This expression implies that k_{des} describes how many molecules are desorbed from a unit area per unit time; if one wants to calculate the time taken for one molecule to desorb from the surface as

$$\tau = \frac{1}{k_{\mathrm{des}}}. \tag{7}$$

τ is a good measure of the mean lifetime of the adsorbate on the surface [12]. Suppose the well-known Arrhenius equation can apply to k_{des}, we obtain

$$k_{\mathrm{des}} = A_{\mathrm{des}}e^{-\frac{E_{\mathrm{des}}}{RT_S}}, \tag{8}$$

where E_{des} is the activation energy associated with the process of desorption and A_{des} is the Arrhenius pre-exponential factor, a.k.a. the frequency factor, whose meaning will become clear soon. One should notice that T_s is not the temperature of gas, but the surface temperature because the adsorbed methane molecule (CH_4*) can be deemed to be a part of the surface, and it is likely to share the same temperature with the surface. To overcome the dissociation barrier, substantial thermal energy has to be provided with the adsorbate from the surface.

In order to clarify what E_{des} and A_{des} mean, it is high time we had a more concrete idea about what would be happening on the surface in question. We are thinking about methane adsorption followed by the dissociation of its C–H bond, as schematically described in Fig. 3a. The corresponding potential energy diagram is shown in Fig. 3b. In this diagram, CH_4(g) represents the state of the gaseous methane molecule, while CH_4* indicates the molecularly adsorbed methane on the surface. Note that the molecularly adsorbed state is not always the same as the physisorbed state. In the latter, the van der Waals (vdw) force dominates the interaction, while in the former, the attractive force resulting from electron exchange between the surface and the adsorbate also makes a contribution more or less on top of the vdw interaction. Generally, we do not assume there is any barrier for the adsorption process. But this is not always the case [13]. Things are not so simple on the surface—so much so that

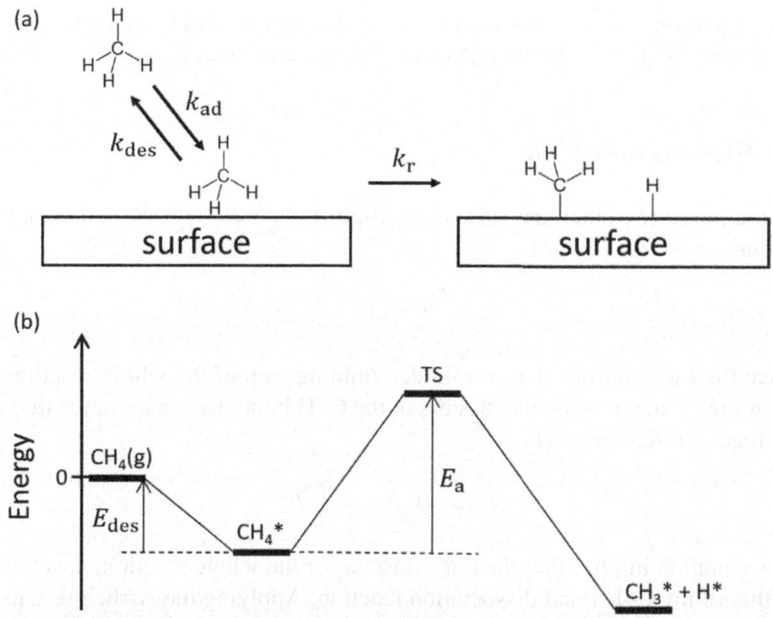

Fig. 3 Schematic representation of the adsorption and the following C–H bond dissociation of methane on a surface (**a**) and the potential energy diagram describing the process indicated in the top panel (**b**). The asterisk mark (*) denotes the adsorbed species on the surface. TS stands for the transition state for the C–H bond dissociation reaction, whose rate constant is denoted as k_r

we need theory and simulation, which can provide us with a vision different from experiment.

In Fig. 3b, the total energy of the system in which gaseous methane is infinitely separated from the surface, denoted as $CH_4(g)$, is set to $E = 0$. The energy difference between $CH_4(g)$ and CH_4*, E_{des}, can be viewed as the activation energy for the desorption of methane from the surface. Inverting both sides of Eq. 8, we have

$$\tau = \tau_0 e^{\frac{E_{des}}{RT_s}}. \tag{9}$$

One should notice that $\tau_0 = \frac{1}{A_{des}}$. This is actually the same as what Frenkel proposed in his seminal work published in 1924 [14]. From Eq. 9, it is clear that τ_0 indicates the minimum lifetime of the adsorbate on the surface at the limit of $E_{des} \to 0$ or $T_s \to \infty$. Hence, it would be reasonable to regard τ_0 as the period of vibration in the direction out of the surface [12].

Now, we are ready to return to the Langmuir adsorption isotherm, Eq. 3, which can be reproduced as follows:

$$\theta = \frac{\frac{k_{ad}}{k_{des}} P}{1 + \frac{k_{ad}}{k_{des}} P}. \tag{10}$$

This equation can be examined by considering two limits: high-pressure and low-pressure conditions. In the followings, we present them in order.

2.1.1 High-Pressure Limit

When the pressure of methane gas is very high, $1 + \frac{k_{\text{ad}}}{k_{\text{des}}} P \approx \frac{k_{\text{ad}}}{k_{\text{des}}} P$, and then, Eq. 10 results in

$$\theta = P. \tag{11}$$

Since the rate-limiting step or rate-determining step of the whole reaction illustrated in Fig. 3 may well be the process of the C–H bond dissociation, the rate of the whole reaction, *R*, then reads

$$R = -k_r\theta = -k_r P. \tag{12}$$

This equation implies that the rate constant for the whole reaction, *k*, is the same as k_r, that of the C–H bond dissociation reaction. Applying the Arrhenius equation, one can have

$$k = k_r = A_r e^{-\frac{E_a}{RT_S}}, \tag{13}$$

where A_r is the Arrhenius pre-exponential factor for the step of the C–H bond cleavage. The Arrhenius plot for this reaction can be justified through.

$$\ln k = \ln A_r - \frac{E_a}{RT_s}. \tag{14}$$

There is nothing interesting here.

2.1.2 Low-Pressure Limit

Let us turn to a situation where the pressure of methane gas is very low so that $1 + \frac{k_{\text{ad}}}{k_{\text{des}}} P \approx 1$; hence, Eq. 10 can be rewritten as

$$\theta = \frac{k_{\text{ad}}}{k_{\text{des}}} P. \tag{15}$$

Again, we make good use of the feature of the rate-determining step, obtaining

$$R = -k_r\theta = -\frac{k_r k_{\text{ad}}}{k_{\text{des}}} P. \tag{16}$$

Then, this equation implies that the rate constant for the whole reaction, k, or the apparent rate constant, should take a form.

$$k = \frac{k_r k_{\rm ad}}{k_{\rm des}}. \tag{17}$$

Since for the description of the rate constants $k_{\rm ad}$, $k_{\rm des}$, and k_r, one can use Eqs. 6, 8, and 13, respectively, we arrive at

$$k = \frac{\tau_0 A_r}{\sqrt{2\pi mkT_g}} e^{-\frac{E_a - E_{\rm des}}{RT_s}}. \tag{18}$$

In this case, the Arrhenius plot is expected to follow

$$\ln k = \ln \frac{\tau_0 A_r}{\sqrt{2\pi mkT_g}} - \frac{E_{\rm app}}{RT_s}, \tag{19}$$

where $E_{\rm app}$ is what is called the apparent activation energy, which reads in this case $E_{\rm app} = E_a - E_{\rm des}$. The formulation of the apparent activation energy becomes more complicated as the kinds of adsorbed species increase [15]. We have to make a note here that the way presented above is not the only way to arrive at the concept of the apparent activation energy, instead one could do so using the so-called steady-state approximation [16].

The magnitude of $E_{\rm des}$ can be viewed as the same as that of adsorption energy. It follows that as the affinity of the surface to methane increases, the apparent activation energy gets smaller. This finding provides us with an important lesson: There is more to the reaction kinetics of methane activation on heterogeneous catalysts than the activation energy of its C–H bond dissociation. Given a lot of experiments have been conducted to scrutinize the activation of methane, whether on pure metals or on metal oxides, in ultra-high-vacuum (UHV) conditions [17–19], we would like to coherently in this chapter stay in the low-pressure regime.

2.2 Two Mechanisms: Direct Mechanism and Trapping-Mediated Mechanism

Generally, there are two mechanisms known for the C–H bond activation of methane: One is referred to as the direct mechanism, and the other one the trapping-mediated mechanism, a.k.a. precursor-mediated mechanism [20–22].

In the direct mechanism, the cleavage of methane's C–H bond happens when the molecule collides onto the surface, which is followed by the concomitant formation of bonds between the H and CH_3 species and the surface atoms. The probability of dissociation strongly depends on the temperature of methane gas, whereas it is not significantly affected by the surface temperature. The kinetic energy of methane is of great importance so that the sufficient energy is transferred to the reaction coordinate, allowing the molecule to surmount the activation barrier. Generally, this mechanism plays an important role when the barrier associated with the C–H bond rapture is substantially large and the adsorption energy of methane to the surface is negligibly small. Thus, it can mainly be applied to the kinetics of methane reaction on pure metal surfaces.

On the other hand, in the trapping-mediated mechanism is first generated the molecularly adsorbed state of methane, i.e., the precursor state. This state can be viewed as part of surface, so it has the same temperature as the surface. It follows that the probability of the C–H bond dissociation in this mechanism depends significantly on the surface temperature. The kinetics competition between the C–H bond breaking reaction and desorption of methane from the surface determines the net reaction rate. Since the reaction begins with the precursor state, it has to have an access to low-barrier paths for dissociation during its lifetime on the surface.

Whether methane on a surface takes the direct mechanism or the trapping-mediated mechanism for its C–H bond dissociation is contingent upon the nature of the molecule–surface interaction. As a general rule of thumb, if the adsorption energy is small and the activation barrier is high, the direct mechanism applies, while if there is a strong molecule–surface interaction and the barrier height is relatively low, then the trapping-mediated mechanism applies. The above-discussed dichotomy is reminiscent of the well-known distinction between the activated and non-activated adsorptions [23].

The difference in energetics between the two mechanisms above is simply depicted in Fig. 4. In the direct mechanism, the desorption barrier is so small that the apparent activation energy, E_{app}, can be regarded as almost the same as that of the true activation energy, E_{a}, or the intrinsic activation energy, while the trapping-mediated mechanism is characterized by a relatively deep initial potential well and the resultant high desorption barrier. Therefore, the deeper the initial potential well becomes, the smaller the apparent activation energy is, and even we have a situation where the apparent activation energy becomes negative due to a substantially large adsorption energy.

Since in the direct mechanism the molecule has to have a substantially large kinetic energy for the reaction, one needs to provide a large thermal energy with the gaseous molecule. However, in such a high-temperature experiment, it would be very difficult to have a control over the subsequent reactions or to deter unfavorable side reactions. This is why, in this chapter, we would like to cling to the trapping-mediated mechanism rather than the direct mechanism. Here, we stop taking a look at the kinetics of methane on the surface, and move on to scrutiny of the energetics of methane on the surface, including the strength of the adsorption and the extent to which the C–H bond of methane is activated in the molecularly adsorbed state.

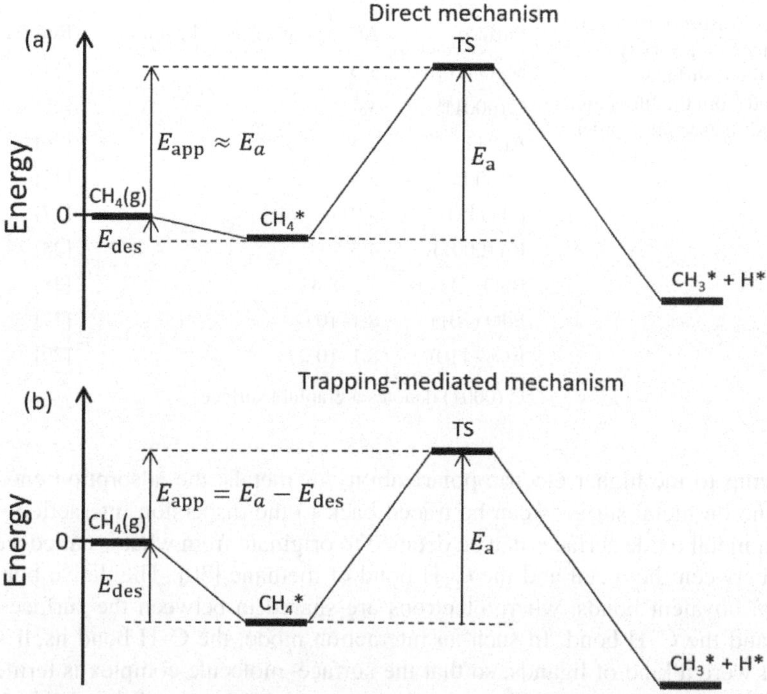

Fig. 4 Typical potential energy diagrams for the C–H bond activation of methane in the direct mechanism (**a**) and trapping-mediated mechanism (**b**)

3 Energetics of Methane on Surface

3.1 How Strongly Methane Can be Adsorbed on the Surface?

In the trapping-mediated mechanism, the adsorption energy of methane on the surface has a decisive role, so in this section, we pause to review the adsorption energy of methane on various surfaces. The development of an optimization-inversion method for analysis of the temperature-programmed desorption (TPD) data has made it possible to obtain the accurate adsorption energy of methane and its desorption-kinetics parameters [24]. Table 1 summarizes the adsorption energies of methane on various surfaces, ranging from the graphite surface to pure metal and metal-oxide surfaces. Some trends can be spotted in this table; for example, generally metal oxides have a larger adsorption energy than the pure metal surfaces. However, the large adsorption energy of methane is not the inherent feature of metal oxides because the MgO (100) surface is characterized by the lowest adsorption energy of the entries of this table. Maybe, being a late transition metal oxide is important for a substantial adsorption energy.

Table 1 Adsorption energies of methane on a variety of well-defined surfaces compiled from the literature, tabulated in ascending order

Surface	Adsorption energy (kcal/mol)	References
MgO (100)	3.0	[24]
C (0001)[a]	3.4	[25]
Au (111)	3.5	[26]
Pt (111)	3.6	[25]
Pd (111)	3.9	[27]
Ru (0001)	4.3	[28]
RuO_2 (110)	7.2–8.8	[28]
PdO (101)	8.1–10.0	[27]
IrO_2 (110)	8.1–10.3	[29]

[a]C (0001) denotes a graphite surface

Owing to the higher electric polarizability of metals, the adsorption energy of methane on metal surfaces can be traced back to the dispersion interaction, while that on metal oxide surfaces can be deemed to originate from what is called a dative bond between the metal and the C–H bond of methane [30]. The dative bond is a kind of covalent bonds, where electrons are shared in between the surface metal atom and the C–H bond. In such an interaction mode, the C–H bond itself serves as if it were a kind of ligands, so that the surface–molecule complex is termed the σ-complex [27]. The detailed electronic aspect of this bond will be clarified later, taking methane–IrO_2 (110) interaction as an example.

From Table 1, it is clear that RuO_2, PdO, and IrO_2 are good candidates for methane activation in the trapping-mediated mechanism. The adsorption and desorption kinetics of methane on these surfaces have well been characterized with the help of ultra-high-vacuum techniques. Their well-defined surface features will be reviewed in the next section.

3.2 *PdO, IrO_2, and RuO_2*

Let us begin with the bulk crystalline structures of PdO [31], IrO_2 [32], and RuO_2 [33]. As shown in Fig. 5, IrO_2 and RuO_2 share the same structure. PdO belongs to a space group of $P4_2/mmc$, while IrO_2 and RuO_2 $P4_2/mnm$. As such, all of them have a tetragonal unit cell. However, the coordination environments of the transition metals and oxides are different between the two unit cells. The crystal structure of PdO is the so-called PtS structure, and square-planar Pd atoms and tetrahedral O atoms are included. As for the IrO_2/RuO_2 structure, they are classified as the famous rutile-type one, wherein the cationic Ir or Ru centers are located at a sixfold-coordinated (octahedral) site, while the O atoms at a threefold-coordinated (trigonal planar) site. In this chapter, periodic structures, such as bulk and surface, are henceforth visualized with the help of VESTA [34].

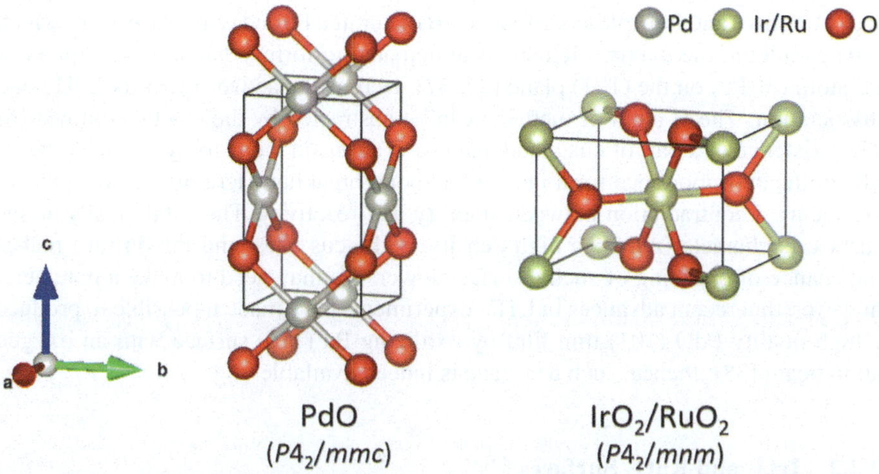

Fig. 5 Crystal structures of PdO, IrO_2, and RuO_2. As for PdO, some atoms and bonds out of the boundary of the unit cell are also shown

3.2.1 PdO Surface

The thermodynamic stability of the PdO surface has been widely investigated, and the PdO (100) orientation has proven to be the most stable surface, dominating its Wulff polyhedron [35]. PdO (101) is the second most stable surface and accounts for a smaller fraction of the surface. The structures of the PdO (100) and (101) surfaces are shown in Fig. 6. In PdO, Pd has the d^8 electron count, making it easier to take a square-planar coordination structure. The favorable coordination number of Pd^{2+} is 4 [36]. Since there is no coordinatively unsaturated (cus) Pd atom on the (100) facet, this surface becomes stable and takes precedence over the (101) plane, which includes cus Pd atoms. On the (101) surface, we can see alternating rows of threefold-coordinated (3f) Pd atoms (cus Pd atoms), threefold-coordinated O atoms

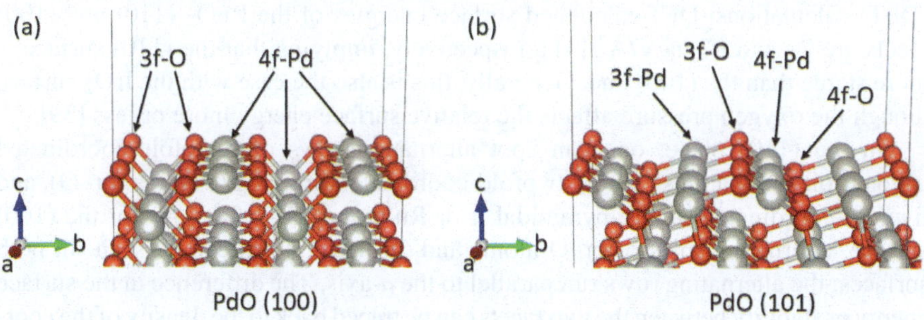

Fig. 6 Structures of the stable PdO surfaces: nonstoichiometric PdO (100) surface (**a**) and stoichiometric PdO (101) surface (**b**). Threefold (fourfold) coordinated O and Pd atoms are denoted as 3f (4f)-O and 3f (4f)-Pd, respectively

(cus O atoms), and fourfold-coordinated (4f) saturated Pd and O atoms, each of which runs parallel to the *a*-axis. Methane is anticipated to form a dative bond with a cus Pd atom (3f-Pd) on the (101) plane [27, 37]. Here is the active site of its C–H bond dissociation. The H atom of methane can be abstracted by the cus O atom nearby. The existence of a pair of cus metal and cus O atoms in the vicinity of each other is of significant importance for a facile C–H bond breaking. Here, we realize that we are facing a contradiction between stability and reactivity. The catalytically active surface is characterized by a high density of the cus sites, and this in turn makes the chance of exposing of such a surface lower. We may need to make a note here, however, that recent advances in UHV experiments have made it possible to produce a high-quality PdO (101) thin film by oxidizing Pd (111) surface with an oxygen atom beam [38]; thence, such a surface is indeed available.

3.2.2 IrO_2 and RuO_2 Surfaces

There are two dominant surfaces for IrO_2 and RuO_2: (110) and (100) facets. Generally, they can be generated by oxidizing the surface of pure metal Ir or Ru. The oxidation of the Ir (111) surface at an oxygen pressure of about 100 mbar and higher temperatures of 775 K or above results in (110)- and (100)-oriented domains of rutile IrO_2 [37]. Their surface characteristics have been well investigated by utilizing in situ surface X-ray diffraction technique. Also, one could generate a high-quality IrO_2 (110) plane by oxidizing Ir (100) surface [29]. By changing oxygen exposure during oxidation, a corundum Ir_2O_3 (001) domain appears [40].

The RuO_2 (110) surface can be grown epitaxially on the Ru(0001) plane during the oxidation of Ru, while the growth of the RuO_2 (100) facet can be observed on the Ru $(10\bar{1}0)$ plane [41]. In contrast with Ir, lattice matching determines the selectivity of either (110) or (100) surfaces in the case of Ru. It is also possible to yield both (110) and (100) facets of IrO_2 or RuO_2 on the (001) surface of $SrTiO_3$ or MgO substrates on the basis of pulsed laser desorption method [42].

Properties of those surfaces have been thoroughly scrutinized by employing low-energy electron diffraction (LEED) techniques as well as density functional theory (DFT) calculations. DFT-calculated surface energies of the RuO_2 (110) and (100) facets are 71 and 87 meV/Å^2 [41], respectively, implying that the (110) surface is more stable than the (100) one. Generally, this is also the case with the IrO_2 surface though the oxygen pressure affects the relative surface energy more or less [39].

On the (110) plane, one can spot alternating rows of threefold-coordinated trigonal-planar O atoms (3f–O), twofold-coordinated bridging O atoms (br–O), and fivefold-coordinated square-pyramidal Ir or Ru atoms (5f–Ir/Ru). As for the (100) surface, alternating rows of br–O atoms and 5f–Ir/Ru atoms can be seen. In both surfaces, the alternating rows run parallel to the *a*-axis. The difference in the surface energy or stability between the two facets can be traced back to the density of the coordinatively unsaturated (cus) sites, namely the 5f–Ir/Ru and br–O sites. The density of the cus sites on the (100) surface is higher than that of the (110) surface.

There may be a discussion here as to whether or not the 3f–O atoms can be regarded as unsaturated sites. In the case of the PdO surface, they are no doubt classified as an unsaturated site because the O atoms are in the tetrahedral coordination environment and they have four bonding partners in the crystalline bulk structure. However, in the case of the IrO_2 or RuO_2 surface, they might well be regarded as saturated because their coordination pattern is not tetrahedral and they enjoy only three bonding partners even in the bulk structure.

Similarly to the (101) surface of PdO, one can easily find a pair of cus metal and cus O atoms in the proximity of each other both on the (110) and (100) surfaces of IrO_2 and RuO_2. The 5f–Ir/Ru sites appear to act as an active site for methane activation. A dative bond between the C–H bond of methane and the surface metal cus site will be generated. However, there is a recent study that uses the low-energy ion scattering spectroscopy (LEISS) techniques, temperature-programmed reaction spectroscopy (TPRS) methods, and density functional theory (DFT) calculations [43], in which the authors found that the cus 5f–Ir sites on the IrO_2 (100) surface are completely terminated by oxygen atoms. The condition of being devoid of the cus Ir atoms on the (100) surface makes it less reactive, so that it may be challenging to prepare a catalytically active IrO_2 (100) surface. This observation can be corroborated by a TPRS spectrum showing a significantly weak binding of H_2O and CO molecules on the surface [43]. Were it possible to generate an O-free, stoichiometric IrO_2 (100) surface, the C–H bond of methane could be effectively activated [20].

When it comes to the RuO_2 (110) surface, two regions with (1×1) and $c(2 \times 2)$ symmetries have been identified, and the (1×1) phase possesses a high catalytic activity, yet the $c(2 \times 2)$ one does not [44]. The coexistence of the (1×1) and $c(2 \times 2)$ phases on the RuO_2 (110) surface makes things intricate. Though a low-energy-electron diffraction inspection supported by a scanning tunneling microscopy (STM) measurement pointed out that the $c(2 \times 2)$ phase is a reconstructed structure [45], the surface structure of the $c(2 \times 2)$ phase has yet been unknown at the atomic level.

As far as the (1×1) phase of the (110) surface of RuO_2 goes, the surface structure has been well-defined and is expected to be close to what is shown in Fig. 7 with the help of LEED and DFT [41]. Its high catalytic activity toward CO is known [44].

3.3 Adsorption of Methane on a Metal Oxide

A modus operandi of methane activation on the surface of a transition metal oxide is the following: First methane's C-H bond forms a dative interaction with a cus metal site on the surface, and second the neighboring cus O site abstracts the H atom of the activated C–H bond [46]. In the literature, one can easily find a variety of experimental and theoretical studies on methane activation on late transition metal oxide surfaces, including PdO (101) [37, 47, 48], IrO_2 (110) [29, 29–51], and RuO_2 (110) surfaces [20, 52]. Among them, the study of the IrO_2 (110) surface [29] has made a significant impact because it was experimentally revealed that the initial C–H bond dissociation of methane highly facilely occurs on the surface at temperatures

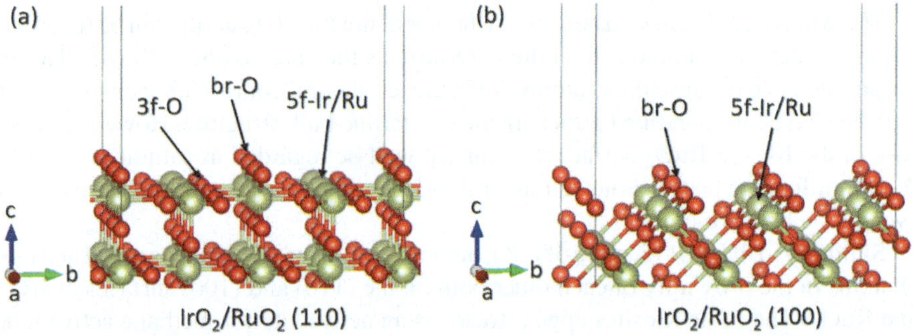

Fig. 7 Structures of the stable stoichiometric IrO_2/RuO_2 surfaces: IrO_2/RuO_2 (110) surface (**a**) and IrO_2/RuO_2 (100) surface (**b**). Threefold-coordinated O atoms are denoted as 3f–O, twofold-coordinated bridging O atoms, br–O, and fivefold-coordinated Ir or Ru atoms, 5f–Ir/Ru

as low as 150 K. Thus, in this section, taking the IrO_2 surface as an example, we shall take a closer look at the adsorption and activation of methane on that surface.

3.3.1 Adsorption Geometry of CH_4 on the IrO_2 (110) Surface

Figure 8 shows the adsorption structure of methane on the IrO_2 (110) surface obtained from a periodic-DFT computation with a functional of Perdew–Burke–Ernzerhof (PBE) [53]. This structure is taken from our recent study; thus, the reader is encouraged to take a look at Ref. [51] for more details of computation.

Methane appears to straddle a cus 5f–Ir atom by opening the H1–C–H2 angle to 122.6°. The opposite H–C–H angle pointing out of the surface has an angle of 112.5°. Note that the normal H–C–H angle in methane is 109.5°. The C–H1 bond is closer to the Ir atom than the C–H2 bond, so the C–H1 bond looks highly activated, for the length of the C–H1 bond is more elongated by 0.05 Å compared to the C–H2

Fig. 8 Adsorption structure of methane on the IrO_2 (110) surface optimized by using a DFT functional of PBE. Selected interatomic distances are shown in units of Å

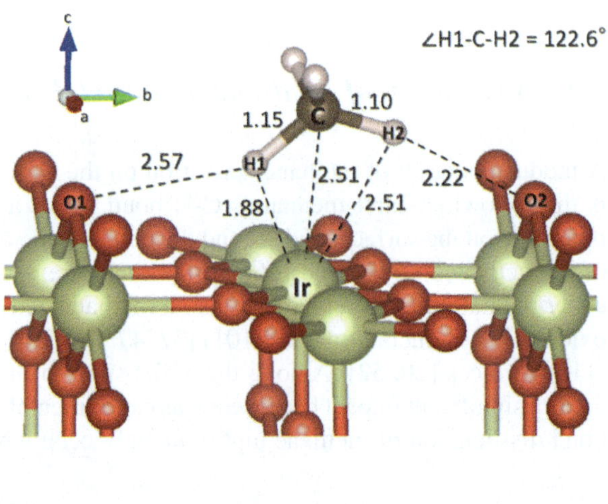

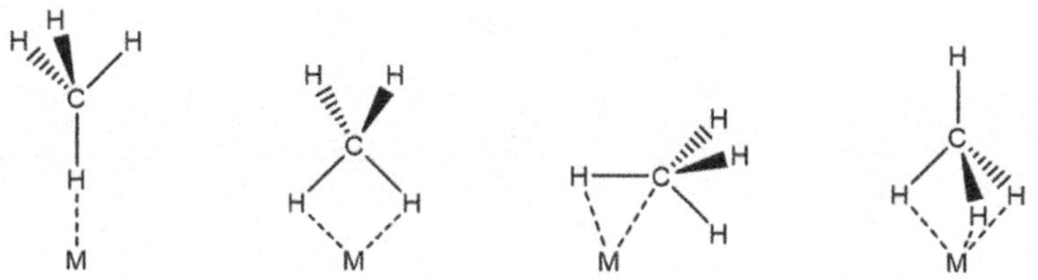

$\eta^1 - H$ $\eta^2 - H, H$ $\eta^2 - C, H$ $\eta^3 - H, H, H$

Fig. 9 Various coordination manners of methane to a metal atom M with notation used in organometallic chemistry

bond. Therefore, the H1 atom is more easily abstracted by the O1 atom than the H2 atom is abstracted by the O2 atom, despite a shorter separation between the H2 and O2 atoms than the H1–O1 distance.

It would be instructive to begin with the terminology that describes the interaction between methane and a cus metal center. Methane can be viewed as a ligand of a certain kind, and so the nomenclature used in organometallic or coordination chemistry is applicable [54]. Some typical coordination modes of methane to a metal cation are schematically illustrated in Fig. 9 with notation describing the interaction. In these coordination modes, the electrons in the $\sigma_{\text{C-H}}$ orbital are donated to an empty orbital of the metal center. Also, the back-donation of electrons from the occupied metal orbital to an unoccupied $\sigma_{\text{C-H}}{}^*$ orbital of methane may play an important role [55]. This is why the complexes shown in Fig. 9 are referred to as σ complex.

Based on the nomenclature shown in Fig. 9, one can judge that the adsorption mode of methane in Fig. 8 can be termed $\eta^2 - C, H$. In this σ complex, therefore, a two-electron three-center bond interaction is expected to play a key role in the C–H bond activation. A quantum chemical aspect of the $\eta^2 - C, H$ interaction will be clarified in the subsequent sections.

3.3.2 A Quantum Chemical Approach to the Bond: COOP

There may be some ways to approach the electronic aspect of the two-electron three-center bond between C–H and Ir. Here we intend to employ a powerful tool named COOP, which is the abbreviation of the crystal orbital overlap population [56, 57], and this can be viewed as a periodic system analogue to Mulliken's overlap population [58].

It would be appropriate here to digress a little bit so as to give a clear idea of COOP to the reader. The definition of COOP for a pair of atoms *i* and *j* can read [59]

$$\mathrm{COOP}_{ij}(E) = S_{ij} \sum_{n,\vec{k}} C^*_{in}\left(\vec{k}\right) C_{jn}\left(\vec{k}\right) \delta\left(\varepsilon_n\left(\vec{k}\right) - E\right), \tag{20}$$

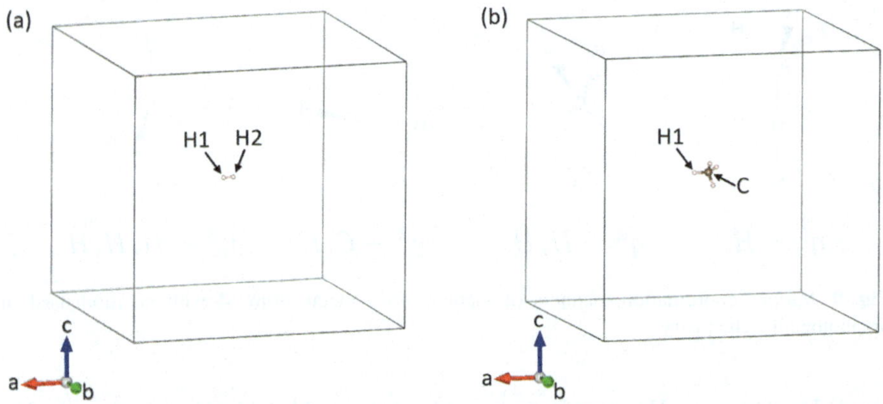

Fig. 10 Model systems used for demonstration of COOP. H_2 (**a**) and CH_4 (**b**) molecules are in a 20 Å × 20 Å × 20 Å cubic box

where S_{ij} denotes the corresponding overlap matrix element for the atom pair, and $C_{jn}\left(\vec{k}\right)$ and $\varepsilon_n\left(\vec{k}\right)$ represent the expansion coefficient of the atom j in the nth crystal orbital at a k-point $\vec{k}$ and its eigenvalue, respectively. Since $\delta\left(\varepsilon_n\left(\vec{k}\right) - E\right)$ corresponds to the density of states (DOS), COOP can be regarded as a weighted DOS. Of course, one can also use orbital indices for i and j instead of those for atoms. From its definition, COOP has a dimension of states/eV.

Let us take a look at some applications of COOP to a molecule in a box (see Fig. 10). As a molecule, adopting H_2 and CH_4 is in order. H_2 is for simplicity, and CH_4 is our object. They are placed at the center of a large periodic box so that the interaction with their replicas in the neighboring cells can be negligible. The H1–H2 and H1–C bonds are aligned parallel to the a (x)-axis.

Figure 11 is the COOP curve for the interaction between the H1 and H2 atoms shown in Fig. 10a. This calculation was carried out by using the extended Hückel molecular orbital (eHMO) method [60] implemented in the YAeHMOP software [61] interfaced with Avogadro 1.2.0 [62–64]. eHMO parameters used for our computation are summarized in a table presented at the end of this chapter.

Understanding the COOP curve is not that difficult. One should bear in mind that the positive COOP value corresponds to a bonding-type interaction, while the negative one an anti-bonding interaction [57]. In a molecular-like calculation, or a molecule-in-a-large-box calculation, the Fermi level is pinned to the energy level of the highest occupied molecular orbital (HOMO) because such an energy point is the boundary between the occupied and unoccupied regions. The positive peak at $E = 0$ can be assigned to the HOMO of H_2, namely the σ orbital. On the other hand, the large negative peak at $E = 22$ eV can be assigned to the LUMO of H_2, or the $\sigma*$ orbital. The bonding and anti-bonding features of the σ orbitals are clearly reflected on the sign of the peaks.

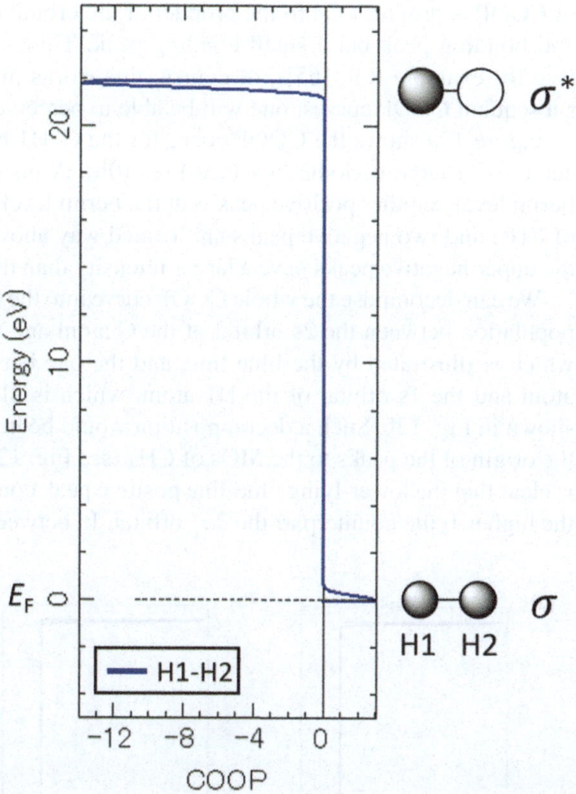

Fig. 11 COOP curve for the interaction between the two H atoms in H_2. The energy step size was set to 0.1 eV, the Gaussian smoothing was used with a standard deviation value of 0.2 eV, and a *k*-point sampling of 1 × 1 × 1 was used. The Fermi level (E_F) is set to zero (horizontal dashed line). Two peaks are assigned to the σ and $\sigma*$ orbitals of the H_2 molecule depicted on the right-hand side

Someone might pose a question here: Why is the intensity of the $\sigma*$-orbital peak is larger than that of the σ-orbital counterpart? The answer can be fetched from a freshman chemistry course. The wave functions for the σ and $\sigma*$ orbitals are written as [65].

$$\psi_\sigma = \frac{1}{\sqrt{2(1 + S_{12})}}(\chi_1 + \chi_2), \tag{21}$$

and

$$\psi_{\sigma^*} = \frac{1}{\sqrt{2(1 - S_{12})}}(\chi_1 - \chi_2), \tag{22}$$

where χ_1 and χ_2 represent the 1s atomic orbital centered on the 1st and 2nd hydrogen atoms, respectively, and S_{12} denotes the overlap between them. Since $S_{12} > 0$, the orbital coefficient for the bonding orbital σ, $1/\sqrt{2(1 + S_{12})}$, is perforce smaller than that for the anti-bonding one $\sigma*$, $1/\sqrt{2(1 - S_{12})}$. As Eq. 20 suggests, the intensity of COOP is proportional to the product of the orbital coefficients, so we see a large anti-bonding peak but a small bonding peak. This orbital feature is a general one (see for example Ref. [65]); of course, this works in the case of CH_4 [66]. In the subsequent COOP curves, one will be able to perceive such a feature.

Figure 12a shows the COOP curve for the C–H1 bond of methane placed at the center of a large periodic box (see Fig. 10b). A positive peak is found below the Fermi level, another positive peak is at the Fermi level, corresponding to the HOMO of CH_4, and two negative peaks are located way above the Fermi level. Notice that the upper negative peaks have a larger intensity than the lower positive counterparts.

We can decompose the whole COOP curve into the contributions from the overlap population between the 2s orbital of the C atom and the 1s orbital of the H1 atom, which is illustrated by the blue line, and the one between the $2p_x$ orbital of the C atom and the 1s orbital of the H1 atom, which is illustrated by the green line, as shown in Fig. 12b. Such a decomposition would be helpful when one wants to trace the origin of the peaks to the MOs of CH_4 (see Fig. 12c). By collating Fig. 12b, c, it is clear that the lower-lying blue-line positive peak comes from the $1a_1$ orbital, while the higher-lying counterpart the $2a_1^*$ orbital. In between these two peaks are located

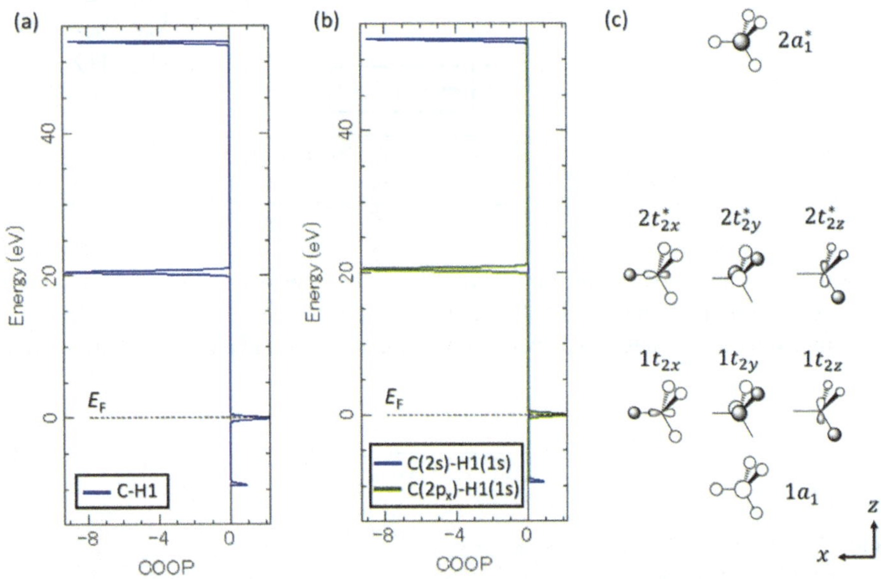

Fig. 12 **a** COOP curve for the interaction between the C and H1 atoms of CH_4 shown in Fig. 10. **b** The whole COOP curve for the C–H1 bond is decomposed into the contributions from the interaction between C′s 2s orbital and H′s 1s orbital (indicated by the blue line) and that between C′s $2p_x$ and H′s 1s orbital (indicated by the green line). **c** Molecular orbitals of CH_4 are delineated on the right-hand side of the COOP graphs so that one can see a correspondence between the COOP peaks and the orbital levels. The energy step size was set to 0.1 eV, the Gaussian smoothing was used with a standard deviation value of 0.2 eV, and a k-point sampling of $1 \times 1 \times 1$ was used. The Fermi level (E_F) is set to zero (horizontal dashed line)

the green-line positive and negative peaks coming, respectively, from the $1t_{2x}$ and $2t^*_{2x}$ orbitals. One should notice that the energy level of the $2a^*_1$ orbital and that of the degenerate $2t^*_2$ set are often switched depending on the methodology used [67].

3.3.3 Effects of Methane's Deformation on the COOP Curve

Next, we turn to the interaction between the orbitals of methane illustrated in Fig. 12c and the surface orbitals of IrO_2. However, as indicated in Fig. 8, we cannot pass by the effects of deformation of methane. In order to look into the COOP curve for the C–H bond of the deformed methane, we take out the deformed methane structure from the $CH_4/IrO_2(110)$ complex and place it into a large periodic box as those shown in Fig. 10. Thus-calculated COOP curve is shown in Fig. 13. Comparing this

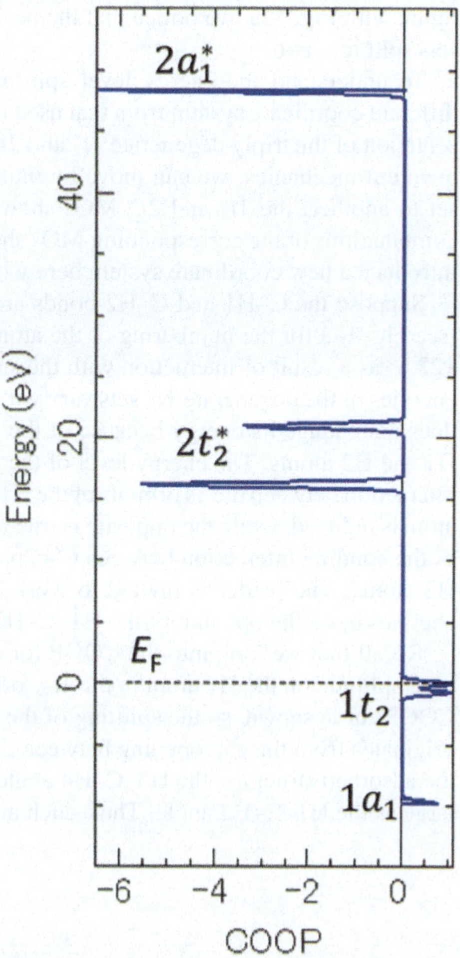

Fig. 13 COOP curve for the C–H1 bond of the deformed CH_4 molecule adsorbed on the IrO_2 (110) surface (see Fig. 8). This molecule is isolated from the surface and placed at the center of a large periodic box in this calculation. The energy step size was set to 0.1 eV, the Gaussian smoothing was used with a standard deviation value of 0.2 eV, and a *k*-point sampling of 1 × 1 × 1 was used. The Fermi level (E_F) is set to zero (horizontal dashed line)

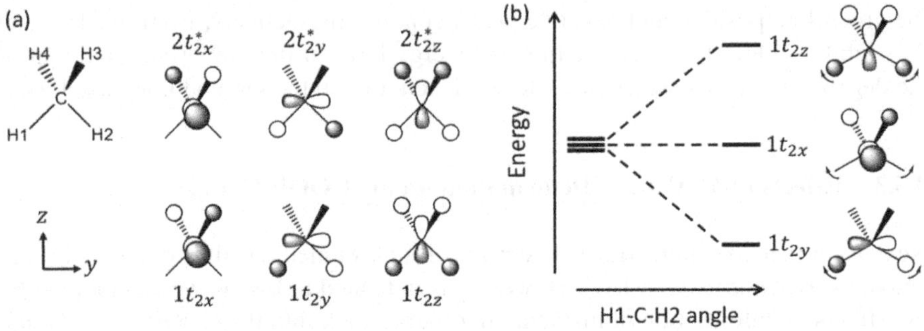

Fig. 14 **a** Alternative representation of the triply degenerate $1t_2$ and $2t_2^*$ manifolds and **b** a schematic representation of how the energy levels of the $1t_2$ set change upon the opening of the H1–C–H2 angle

figure with Fig. 12a, we notice that the peaks ascribed to the $1t_2$ and $2t_2^*$ degenerate sets split into two.

To understand the energy level splitting due to the deformation, we adopt a different coordinate system from that used in Fig. 12c, depicting an alternative representation of the triply degenerate $1t_2$ and $2t_2^*$ orbitals. Owing to the fundamentals of quantum mechanics, we can move by unitary transformation from one degenerate set to another; the $1t_2$ and $2t_2^*$ MOs shown in Fig. 14 can be expressed as linear combinations of the corresponding MOs shown in Fig. 12c [65]. The reason why we introduce a new coordinate system here will be clear soon.

Suppose the C–H1 and C–H2 bonds are directed toward the IrO_2 (110) surface (see Fig. 14a for the numbering of the atoms), the H1–C–H2 angle is opened up to 122.6° as a result of interaction with the surface. Figure 14b shows how the orbital energies of the degenerate $1t_2$ sets vary with change in the H1–C–H2 angle. The $1t_{2x}$ does not change its energy because of the absence of the orbital amplitudes on the H1 and H2 atoms. The energy level of the $1t_{2z}$ orbital goes up because the bonding interaction between the 1s orbitals of the H1 and H2 atoms and the $2p_z$ orbital of the C atom is reduced, while the opposite is true for the t_{2y} orbital due to the reinforcement of the bonding interaction between C′s $2p_y$ orbital and the 1s orbitals of the H1 and H2 atoms. The reader is invited to work out how the $2t_2^*$ manifolds change their energies upon the opening of the H1–C–H2 angle.

Recall that we look into the COOP for the C–H1 bond in Fig. 13. Since there is no amplitude on the H1 atom in the $1t_{2x}$ orbital, this orbital cannot contribute to the COOP curve shown, so the splitting of the COOP peak corresponding to the $1t_2$ set originates from the gap opening between the $1t_{2z}$ and $1t_{2y}$ orbitals (see Fig. 14b). In the adsorbed structure, the H3–C–H4 angle is also opened up, but its angle is not as large as the H1–C–H2 angle. Thus, such an effect would be negligible.

3.3.4 Interaction Between Methane's C–H Bond and IrO_2 Surface

Now that we have understood the origin of the peaks in the COOP curve for the deformed methane molecule, which has the same structure as the one adsorbed on the surface, we are ready to look into the interaction between the C–H bond and the surface orbitals. First, the deformed CH_4 molecule is placed at the middle of the vacuum slab on the IrO_2 (110) surface (see Fig. 15a), and then the COOP curve for its C–H bond is calculated. The shape of the COOP spectrum is generally the same as that shown in Fig. 13 because in this calculation there is no interaction between the C–H bond and the surface, either. However, the location of the Fermi level is different between the two calculations. In Fig. 13, the Fermi level is situated at the same energy location as that of the HOMO of methane, while in the case of Fig. 15a the Fermi level coincides with that of the IrO_2 surface slab. From this figure, it is clear that the methane's HOMOs are so close to the Fermi level that it would be reasonable to assume they would become active orbitals for the adsorption of methane and the following activation of its C–H bond.

In Fig. 15b is shown the COOP curve for the C–H bond of methane which is strongly bound by the IrO_2 (110) surface. Comparing this COOP profile with that shown in Fig. 15a, we see the peaks far from the Fermi level appear to remain

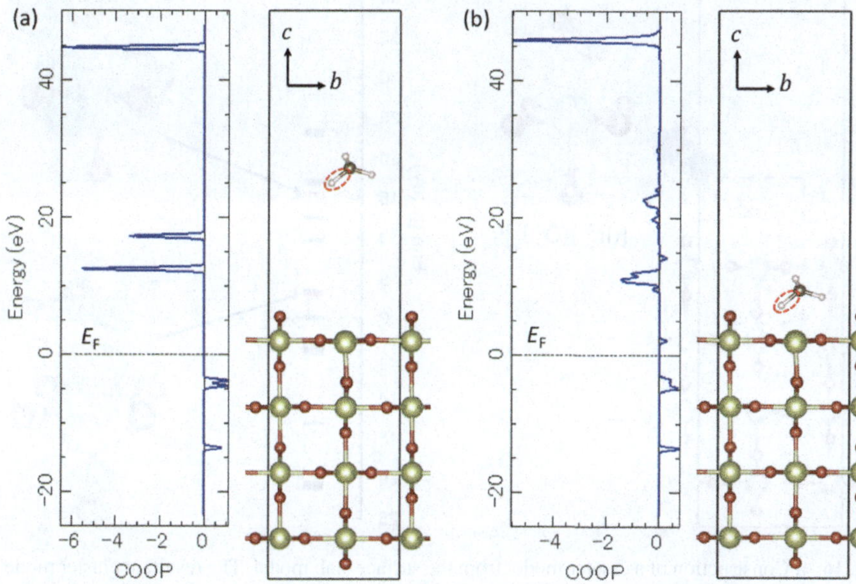

Fig. 15 COOP curve for the C–H bond of a deformed CH_4 molecule (highlighted by the red dashed circle in the structure shown on the right-hand side of each COOP graph). In **a** the deformed CH_4 molecule is placed at the middle of the vacuum slab so that the interaction with the surface can be negligible, while in **b** the methane-adsorbed structure is used for the COOP calculation. The energy step size was set to 0.1 eV, the Gaussian smoothing was used with a standard deviation value of 0.2 eV, and a k-point sampling of $4 \times 6 \times 1$ was used. The Fermi level (E_F) is set to zero (horizontal dashed line)

unchanged. By contrast, around the Fermi level, some changes can be discerned; for example, the peaks get more broadened, and two positive COOP peaks appear above the Fermi level.

In the COOP profile of the C–H bond of the methane molecule without interaction with the surface, positive COOP peaks lie below the Fermi level, while negative ones lie above the Fermi level (see Fig. 15a). This follows that bonding-type C–H orbitals are fully occupied, yet anti-bonding-type C–H orbitals are emptied. However, the interaction with the surface makes a difference: two unoccupied bonding-type C–H orbitals emerge above the Fermi level. This implies that the adsorption of methane makes the C–H bond weak more or less. We wish to identify these peaks, but we are unable to visualize the crystal orbitals, so we make a comparison between cluster-model MOs and the COOP peaks to pursue the origin of the two unoccupied positive COOP peaks.

Figure 16a shows how to construct a cluster model from the surface slab model. Our cluster model is a minimum one, which just includes the deformed CH_4 molecule

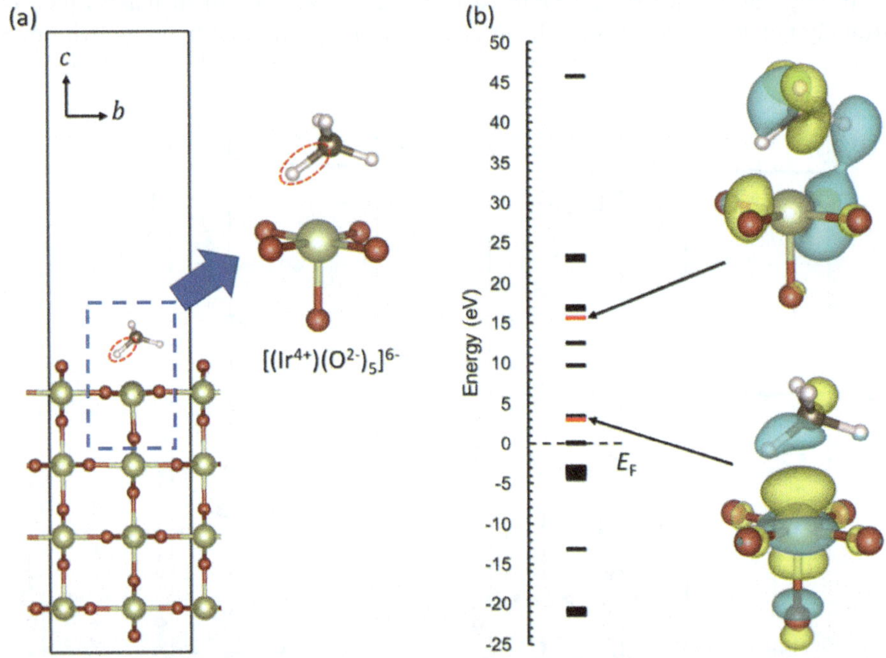

Fig. 16 **a** Construction of a cluster model from the surface slab model. The resultant cluster model consists of the deformed methane molecule adsorbed on the coordinatively unsaturated Ir cation surrounded by five oxide ions. The total charge of the cluster model is set to −6 in accordance with the electron count shown below the model. **b** The MO energy spectrum for the cluster model at the eHMO level of theory. The Fermi level (E_F), which coincides with the HOMO level, is set to zero (horizontal dashed line). The two positive COOP peaks above the Fermi level (see Fig. 15b) are assigned to the MO levels highlighted by red. Their orbital distributions are also shown (the isosurface value is set to 0.05 $e^{1/2}/bohr^{3/2}$)

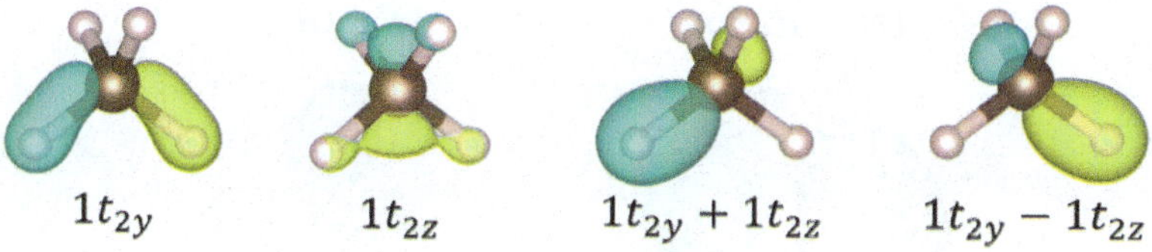

Fig. 17 Isosurface plots of the $1t_{2y}$ and $1t_{2z}$ orbitals and their linear combinations

and the cus Ir atom beneath the CH_4 molecule with the neighboring O atoms. The total charge of the cluster model is set to −6 on the basis of electron count shown below the cluster model in Fig. 16a. Figure 16b shows the calculated MO spectrum of the cluster model at the eHMO level. The two levels highlighted by red are the orbitals which are ascribed to the two positive COOP peaks above the Fermi level in Fig. 15b. Their orbital distributions are also shown. The orbital energies of these levels are consistent with those of the COOP peaks.

It is safe to say that the lower level of the two assigned MOs is generated as a result of an anti-bonding interaction between the σ_{CH} orbital (the HOMO of methane) and the d_{z2} orbital of Ir. Thus, the lobe of the d_{z2} orbital pointing toward CH_4 has the opposite sign to the lobe of the σ_{CH} orbital. On the other hand, as for the upper orbital, in fact, this does not come from the σ_{CH} orbital but from the 2p orbital on the C atom, one of whose two lobes penetrates onto the H atom so that the overlap population between them can take a positive value. As such, the latter orbital interaction is not likely to contribute to the C–H bond activation, so we exclude it from our scope of consideration henceforth.

One may have difficulty correlating the σ_{CH} orbital shown in Fig. 16b with those shown in Fig. 14a. Here one should recall that any linear combination of degenerate eigenfunctions of the Hamiltonian is also an eigenfunction. So we can make the σ_{CH} orbital by taking a linear combination of the $1t_{2y}$ and $1t_{2z}$ orbitals as shown in Fig. 17. The $1t_{2y} + 1t_{2z}$ orbital is the σ_{CH} orbital which interacts with the d_{z2} orbital of Ir.

As the COOP profile shown in Fig. 15b suggests, we can see orbitals generated from a bonding-type interaction between the σ_{CH} orbital and Ir's d orbital in a range from −5 to −3 eV. As shown in Fig. 18, one of such orbitals appears to be formed from the $1t_{2y}$ orbital of methane and the d_{yz} orbital of Ir, while the other one from $1t_{2z}$ and d_{z2}.

A direct way to examine the interaction between the C–H bond and the surface is to look at the COOP profile calculated for the pair of the C atom of methane and the Ir atom on the surface (see Fig. 19a). A large positive C–Ir COOP peak can be observed in a range from −5 to −3 eV, which is immediately rationalized by looking at the orbital interactions depicted in Fig. 18. We see some large negative COOP peaks, but they are located above the Fermi level, so they do not affect the good affinity of Ir toward methane. The total COOP curve for the C–Ir interaction can be decomposed into the contributions from the interactions between the atomic orbitals of the C and Ir atoms. As we have already looked at, the main contributor to the C–Ir COOP is the interaction between the $2p_z$ orbital of C and the $5d_{z2}$ orbital of Ir (see Fig. 19b).

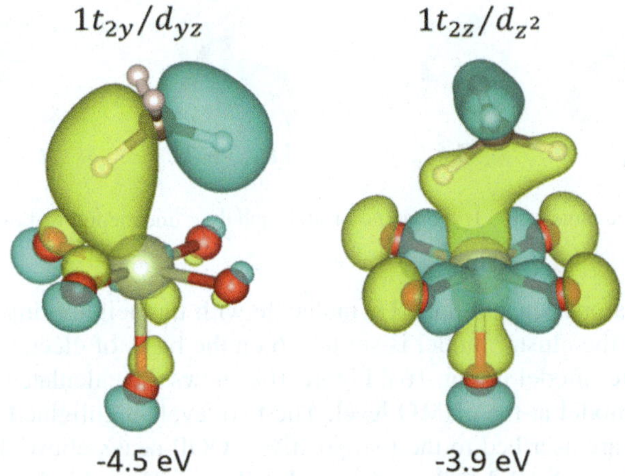

Fig. 18 Isosurface plots of the cluster-model MOs, in which the σ_{CH} orbital interacts with Ir′s *d* orbital in an in-phase manner

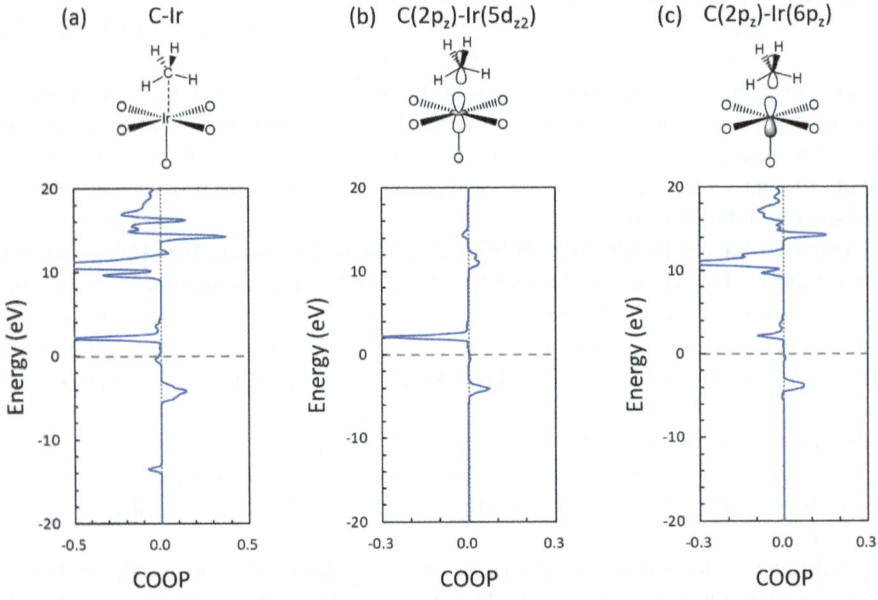

Fig. 19 **a** COOP curve calculated for the C atom of methane and the Ir atom on the surface. The C–Ir COOP profile is decomposed into the contributions from the interaction between C′s $2p_z$ and Ir′s $5d_{z2}$ orbitals (**b**) and the one between C′s $2p_z$ and Ir′s $6p_z$ orbitals (**c**)

Large positive and negative peaks can be found just below and above the Fermi level, respectively, which is a signature of a strong, attractive interaction. The interaction between the $2p_z$ orbital of C and the $6p_z$ orbital of Ir is also found important (see Fig. 19c), but the $6p_z$ orbital is located higher in energy, and so the interaction mainly happens in a region way above the Fermi level. However, one should notice that these orbitals appear to mix into the orbitals generated from the interaction between the $2p_z$ orbital of C and the $5d_{z2}$ orbital of Ir in a second-order fashion. One may wonder if the interaction between C′s $2p_x$ and Ir′s $5d_{xz}$ or between C′s $2p_y$ and Ir′s $5d_{yz}$ would be of importance. In fact, we have checked them, and found that their contribution to the total COOP curve is almost a half of the C($2p_z$)-Ir($5d_{z2}$) and C($2p_z$)-Ir($6p_z$) contributions.

An alternative way to probe the interaction between the σ_{CH} orbital and the surface Ir orbitals is to take a look at the COOP profile calculated for the pair of the Ir and H atoms (see Fig. 20a). This COOP curve looks similar to the one for the C–Ir COOP curve. We observe positive bonding peaks in a range from −3 eV to −5 eV, while the anti-bonding counterpart can be found at around $E = 2$ eV. The decomposition of the total COOP curve into the atomic-orbital contributions, as shown in Fig. 20b–d, reveals that the main contributor is the interaction between H′s 1s and Ir′s $6p_z$ orbitals, yet their contribution is a second-order manner. Also, the contributions from Ir′s $5d_{z2}$ and $5d_{yz}$ orbitals are found important, as can be rationalized by the schematic drawing of the orbital interaction pattern depicted above each COOP plot.

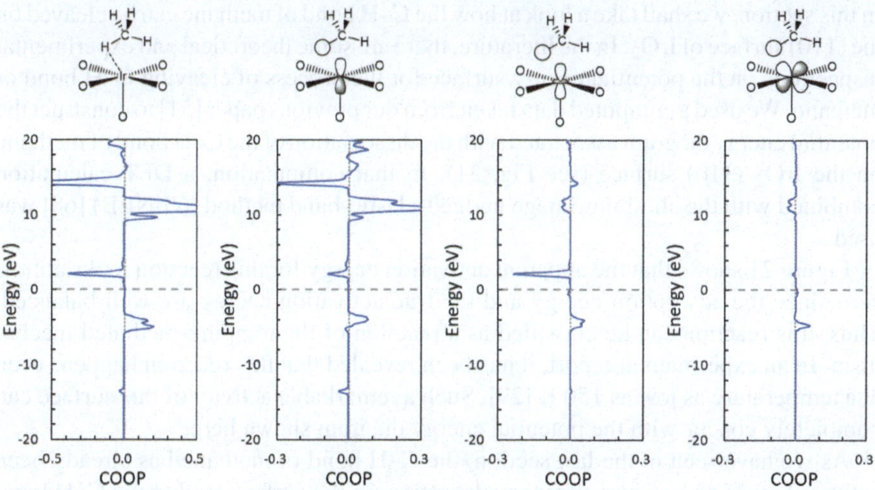

Fig. 20 **a** COOP curve calculated for the H atom of methane and the Ir atom on the surface. The H–Ir COOP profile is decomposed into the contributions from the interaction between H′s 1s and Ir′s $6p_z$ orbitals (**b**), the one between H′s 1s and Ir′s $5d_{z2}$ orbitals (**c**), and the one between H′s 1s and Ir′s $5d_{yz}$ orbitals

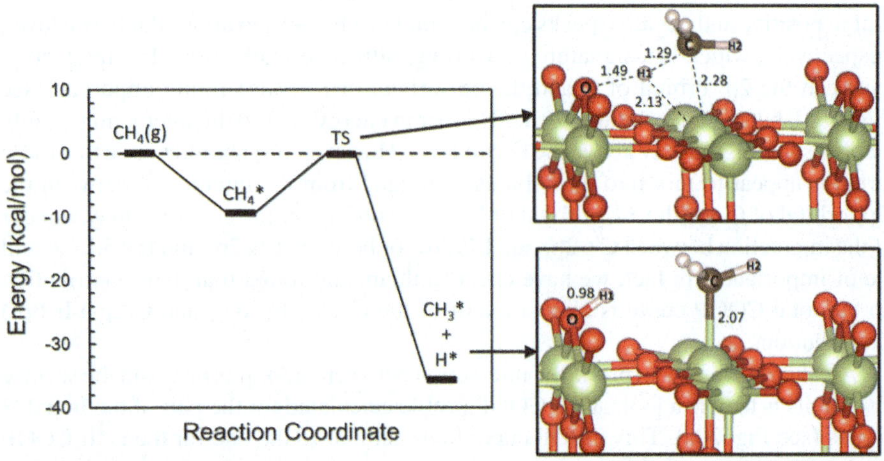

Fig. 21 Computed potential energy diagram associated with the dissociation of the C–H bond of methane on the (110) surface of IrO_2, which has been reported in Ref. [51]. Insets indicate the structures of the transition state (TS) and the final state of the reaction. Selected interatomic distances are shown in units of Å

3.4 C–H Bond Dissociation of Methane on a Metal Oxide Surface

In this section, we shall take a look at how the C–H bond of methane can be cleaved on the (110) surface of IrO_2. In the literature, there are some theoretical and experimental inspections on the potential energy surface for the process of cleaving C–H bond of methane. We used a computed data taken from our previous paper [51] to construct the potential energy diagram associated with the dissociation of the C–H bond of methane on the IrO_2 (110) surface (see Fig. 21). In that computation, a DFT calculation combined with the climbing-image nudged-elastic-band method (CI-NEB) [68] was used.

Figure 21 shows that the apparent activation energy for this reaction looks almost zero since the adsorption energy and the true activation energy are well balanced. Thus, this reaction can be classified as a reaction of the trapping-mediated mechanism. In an experimental report, it has been revealed that this reaction happens even at a temperature as low as 150 K [29]. Such a remarkable activity of this surface can completely concur with the potential energy diagram shown here.

As we have seen in the last section, the C–H bond of methane has already been activated more or less just after the adsorption on the surface so that the C–H bond breaking happens with a relatively low true activation energy. On top of that, the stabilization of the H atom and the CH_3 group in the transition state may also contribute to the lowering of the activation energy. Therefore, it is natural that we should take a closer look at the interaction between the O atom and the abstracted H atom and the one between the C atom and the cus Ir atom.

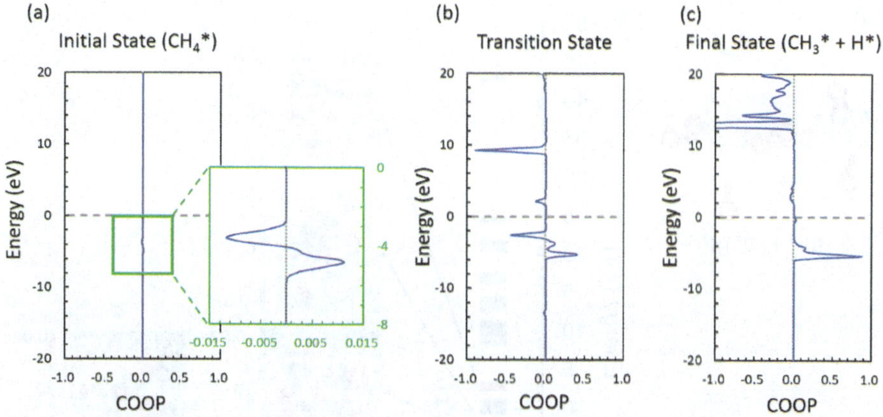

Fig. 22 COOP curve of the interaction between the abstracted H atom and the O atom, which abstracts the H atom, calculated for the initial state (**a**), transition state (**b**), and the final state (**c**) of the C–H bond dissociation reaction. Since the intensity of the COOP peaks for the initial state is very low, a magnified view of a region below the Fermi level is shown

Figure 22 shows how the COOP profile for the interaction between the O and H atoms changes as the reaction of the H-atom abstraction by the O atom proceeds. At the beginning, there are both positive and negative peaks just below the Fermi level, and their energy difference is very small, implying that the overlap between orbitals on the H and O atoms should be very small. Recall that the gap splitting between the bonding and anti-bonding levels as a result of orbital interaction reflects the strength of the interaction. Also, the small intensity of the peaks justifies our expectation. The contribution from the attractive positive peak to the stabilization of the O–H bond is annulled by the repulsive negative peak just above it at the very beginning of the reaction.

As the reaction goes on, the lowest positive COOP peak for the O–H bond gets larger, implying the O–H bonding interaction becomes more and more reinforced. On the other hand, the O–H anti-bonding peak gets pushed up and eventually goes beyond the Fermi level so that the anti-bonding σ_{OH}* orbital can be emptied out. In the transition state, we can perceive the growth of a new bonding peak at around $E = -3$ eV. Its anti-bonding counterpart can also be spotted at around $E = 2$ eV. One should notice that the newly generated anti-bonding peak is located above the Fermi level and this does not affect the stabilization effect brought about by the newly generated bonding COOP peak. Hence, the O–H bond is somewhat already stabilized in the transition state. In the final state, no anti-bonding peak can be seen below the Fermi level, indicating that a robust O–H bond has been formed.

In order to scrutinize the bonding and anti-bonding peaks around the Fermi level found in the COOP profile for the transition state shown in Fig. 22b, we here again exploit the cluster-model calculation. In the cluster model presented in Fig. 16a is not enough for our purpose, for the O atom, which abstract the H atom, is not included. We expand the cluster model used and construct what is shown in Fig. 23a. This

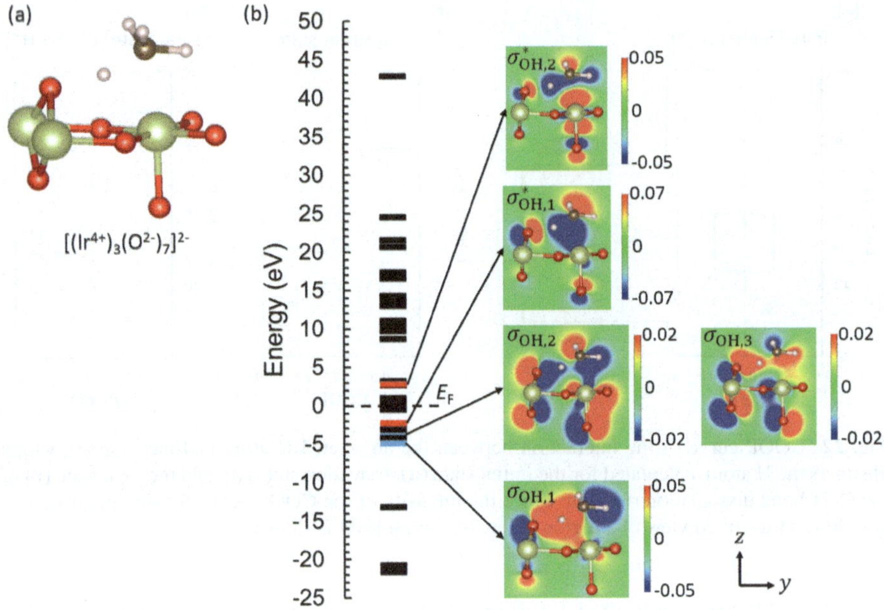

Fig. 23 **a** Construction of a cluster model that can describe orbital interactions in the transition state and **b** its MO energy spectrum calculated at the eHMO level of theory. Contour maps for the MOs, which are assigned to the peaks in the O–H COOP curve in the transition state, are also shown. The blue-colored bands correspond to the orbital where a bonding-type O–H interaction can be seen, while the red-colored ones correspond to the orbital where an anti-bonding-type O–H interaction can be seen. The Fermi level (E_F), which coincides with the HOMO level, is set to zero (horizontal dashed line)

cluster model is constructed from the slab-model structure obtained from the CI-NEB calculation. On the basis of electron count shown below the structure, the total charge of the model is set to −2 and the calculated HOMO energy is regarded as the Fermi level.

In Fig. 23b, one can see the MO energy spectrum for the cluster model shown in Fig. 23a, which was calculated at the level of the eHMO theory. The blue-colored bands correspond to the MOs where a bonding interaction can be found in between the O and H atoms. On the other hand, the red-colored ones correspond to those for the O–H anti-bonding orbitals. Their orbital energy levels are perfectly consistent with the peak positions in the COOP spectrum shown in Fig. 22b. One can perceive that all MO contour maps shown in Fig. 23b include the $1t_2$ manifold or the HOMOs of methane. A bonding interaction between the $2p_y$ orbital of the O atom and each of the $1t_2$ orbitals results in the lowering of the MO energies. Especially, in the $\sigma_{OH,1}$ orbital, the bonding-type O–H–C bond is further stabilized by its in-phase combination with the $5d_{yz}$ orbital of Ir. This orbital interaction seems important for lowering the activation energy by stabilizing the transition state.

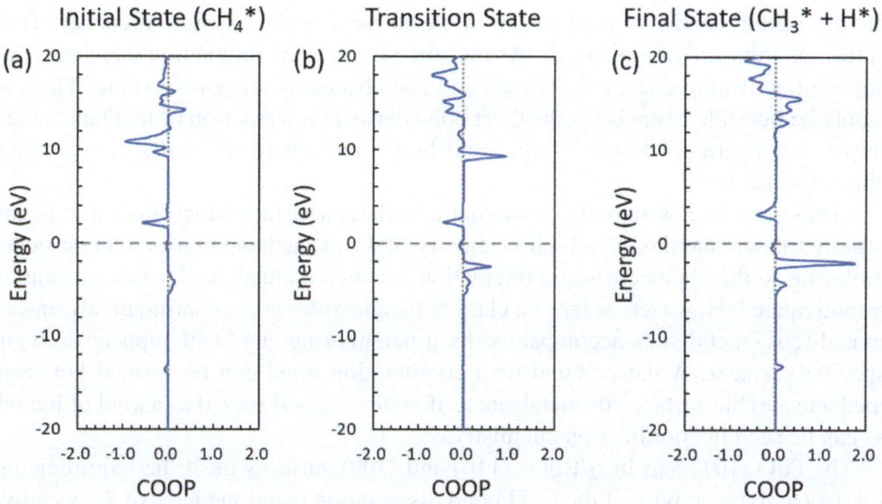

Fig. 24 COOP curve of the interaction between the C atom and the surface cus Ir atom calculated for the initial state **a**, transition state **b**, and the final state **c** of the C–H bond dissociation reaction

Let us turn to the C–Ir bond formation in the C–H bond dissociation reaction. Figure 24 shows how the COOP spectrum calculated for the pair of the C and Ir atoms changes as the reaction proceeds. At the beginning, as we have already looked at in Fig. 19, there is substantial bonding-type interaction between them below the Fermi level. In the transition-state structure, a large positive peak appears at around $E = -2$ eV. Though a tiny anti-bonding peak remains unchanged at the Fermi level even in the TS structure, the new large bonding peak is likely to dominate the Ir–C interaction, and eventually it grows to a huge peak in the final-state structure.

Given the energy location of the newly emerging positive COOP peak, it may be reasonable to correlate this peak with the cluster-model MO, $\sigma^*_{\mathrm{OH},1}$, shown in Fig. 23b. This orbital can be viewed primarily as an anti-bonding OH orbital, yet it can also be viewed as a C–Ir bonding orbital which is pushed up due to its out-of-phase combination with the $2p_y$ orbital of the O atom. From the point of view of the stabilization of the newly formed O–H bond, this orbital interaction seems unfavorable, but this is favored in terms of the stabilization or strengthening of the newly formed C–Ir bond. In this regard, this orbital interaction also plays a key role in making the activation energy lower.

4 Conclusions and Outlook

Given a recent surge of experimental studies utilizing the ultra-high-vacuum technique, it would be important to understand the reaction dynamics of methane on the surface in the low-pressure limit, where the reaction rate of the C–H bond dissociation of methane is governed by a balance between the true activation energy and the desorption barrier of methane from the surface, or the adsorption energy. This is the apparent activation energy. As the surface–methane interaction increases, the apparent activation energy decreases, and even becomes a negative value. Thus, it would be possible to envisage the C–H bond dissociation reaction of methane under the room temperature despite a large bond dissociation energy of methane, exceeding about 100 kcal/mol.

In this context, it is natural that we should seek for a surface which has a significant affinity to methane though a high symmetry of T_d of methane makes it a non-polar molecule so that an electrostatic interaction between methane and a surface cannot be anticipated. However, as far as a class of metal oxides goes, coordinatively unsaturated (cus) metal sites accompanied by a neighboring cus O site appear on their specific surfaces. A dative bond or a coordination bond can be formed between methane and the surface cus metal site as if methane itself served as a kind of ligand as can be seen in coordination chemistry.

The PdO (101) and IrO_2/RuO_2 (110) and (100) surfaces meet the requirements for the catalytic activity of the C–H bond dissociation mentioned above. Especially, the IrO_2 (110) surface shows a remarkable activity, which is exemplified by a low temperature activation of methane thereon. A recent experimental work has reported that the reaction happens at the temperature as low as 150 K [29]. Such a low temperature activation would pave a new way of upgrading methane as an important feedstock for the chemical industry because one could have a control over the reaction there, deterring unfavorable side reactions.

This is why we focus on the activation of the C–H bond of methane on the IrO_2 (110) surface in the latter part of this chapter. A periodic-boundary slab-model calculation is suitable for simulating the surface reaction. However, such a periodicity perforce obscures a discrete nature of the methane molecule on the surface, whose electronic structure should be better described by using the molecular orbital rather than the crystal orbital. Nevertheless, with the help of a band decomposition technique named crystal orbital overlap population (COOP) and some small cluster-model computations, we have gained a clear picture of how methane is adsorbed and activated on the IrO_2 surface in terms of the orbital interaction. Two important features should be noted: First, there are bonding-type orbital interactions between methane's HOMOs and cus Ir's d orbitals, reinforcing the affinity of the surface to methane. Second, in the transition state of the abstraction of the H atom by the cus O atom, the H atom is stabilized by an in-phase orbital interaction among H′s 1s orbital, O′s 2p orbital, C′s 2p orbital, and Ir′s $5d_{yz}$ orbital, resulting in lowering of the activation energy of the C–H bond dissociation reaction.

Enhancing the affinity of the surface to methane is a rational strategy to create a new heterogeneous catalyst that can effectively attract methane and break its robust C–H bond. Thus, theoreticians are now active in finding a surface satisfying such a requirement in order to identify a catalyst that is expected to be more active than the IrO_2 surface. One such example is the (110) surface of β-PtO_2 [51], where a good energy level alignment between methane's HOMO and the unoccupied $5d_{z2}$ orbital of the surface cus Pt atom is achieved. Even if a Pt atom is doped on the rutile-type

TiO_2 (110) surface, one can anticipate a good catalytic activity thereon [69], though TiO_2 per se is not an active catalyst for the C–H bond dissociation.

The C–H bond scission of methane is one of the most important yet challenging steps in methane conversion. But this is not the goal. For example, C–O or C–C bond formation is necessary for the utilization of methane as a chemical building block or chemical raw material. Indeed, it is necessary that CH_3 species be strongly bound by the surface cus metal site when it comes to making the activation energy of the first C–H bond dissociation of methane lower, but at the same time such a strong binding feature would make it difficult for CH_3 to become mobile on the surface. The activation energy for subsequent reactions would perforce become high. The generated CH_3 species may well be doomed to over oxidation. A breakthrough to cut the Gordian knot is really needed.

Acknowledgements This work was supported by KAKENHI grants (numbers JP17K14440 and JP17H03117) from the Japan Society for the Promotion of Science (JSPS) and the Ministry of Education, Culture, Sports, Science and Technology of Japan (MEXT) through the MEXT projects Integrated Research Consortium on Chemical Sciences, Cooperative Research Program of Network Joint Research Center for Materials and Devices and Elements Strategy Initiative to Form Core Research Center, and by JST-CREST JPMJCR15P5 and JST-Mirai JPMJMI18A2. The computations in this work were primarily performed using the computer facilities at the Research Institute for Information Technology, Kyushu University. Y.T. is grateful for a JSPS Grant-in-Aid for Scientific Research on Innovative Areas (Discrete Geometric Analysis for Materials Design, grant numbers JP18H04488 and JP20H04643, and Mixed Anion, grant number JP19H04700).

Appendix

EHMO parameters used for Ir, C, Cl, and O appear in Table 2. They are well-accepted values used in previous studies [70, 71].

Table 2 Extended Hückel parameters

Orbital	H_{ii} (eV)	ζ_1	(c_1)	ζ_2	(c_2)
Ir 6s	−11.36	2.5			
Ir 6p	−4.5	2.2			
Ir 5d	−12.17	5.796	(0.6351)	2.557	(0.5556)
O 2s	−32.3	2.275			
O 2p	−14.8	2.275			
C 2s	−21.4	1.625			
C 2p	−11.4	1.625			
H 1s	−13.6	1.300			

References

1. Kvenvolden K (1988) Chem Geol 71: 41–51
2. Howarth RW, Santoro R, Ingraffea A (2011) Climatic Change 106: 679–690
3. Caballero A, Perez PJ (2013) Chem Soc Rev 42: 8809–8820
4. Shilov AE, Shul'pin GB (1987) Russ Chem Rev 56: 442–464
5. Horn R, Schlögl R (2015) Catal Lett 145: 23–39
6. Langmuir L (1918) J Am Chem Soc 40: 1361–1403
7. Ruthven DM (1984) Principles of Adsorption and Adsorption Processes. John Wiley & Sons, New York
8. Hertz H (1882) Ann Phys 17: 178–193
9. Knudsen M (1915) Ann Phys 47: 697–705
10. Herman MA, Richter W, Sitter H (2004) Epitaxy: Physical Principles and Technical Implementation. Springer, New York
11. L'vov BV (2007) Thermal Decomposition of Solids and Melts: New Thermochemical Approach to the Mechanism, Kinetics and Methodology. Springer, New York
12. Dewing J, Robertson AJB (1957) Proc Roy Soc A240: 423–436
13. Gazdzicki P, Uvdal P, Jakob P (2009) J Chem Phys 130: 224703
14. Frenkel J (1924) Z Phys 26: 117–138
15. Adamson AW (1976) The Physical Chemistry of Surfaces. John Wiley & Sons, New York
16. Sun YK, Weinberg WH (1990) J Vac Sci Technol A 8: 2445–2448
17. Hellman A, Resta A, Martin NM, Gustafson J, Trinchero A, Carlsson PA, Balmes O, Felici R, van Rijn R, Frenken JWM, Andersen JN, Lundgren E, Gronbeck H (2012) J Phys Chem Lett 3: 678–682
18. Zaera F (1991) Catal Lett 11: 95–104
19. Abild-Pedersen F, Lytken O, Engbaek J, Nielsen G, Chorkendorff I, Nørskov JK (2005) Surf Sci 590: 127–137
20. Weaver JF, Hakanoglu C, Antony A, Asthagiri A (2014) Chem Soc Rev 43: 7536–7547
21. Seets DC, Wheeler MC, Mullins CB (1997) Chem Phys Lett 266: 431–436
22. Mullins CB, Weinberg WH (1990) J Chem Phys 92: 4508–4512
23. Kolasinski KW (2008) Surface Science: Foundations of Catalysis and Nanoscience. John Wiley & Sons Ltd., Chichester
24. Tait SL, Dohnalek Z, Campbell CT, Kay BD (2005) J Chem Phys 122: 164708
25. Tait SL, Dohnalek Z, Campbell CT, Kay BD (2006) J Chem Phys 125: 234308
26. Lavrich DJ, Wetterer SM, Bernasek SL, Scoles G (1998) J Phys Chem B 102: 3456–3465
27. Weaver JF, Hakanoglu C, Hawkins JM, Asthagiri A (2010) J Chem Phys 132: 024709
28. Li T, Kim M, Rai R, Liang Z, Asthagiri A, Weaver JF (2016) Phys Chem Chem Phys 18: 22647–22660
29. Liang Z, Li T, Kim M, Asthagiri A, Weaver JF (2017) Science 356: 299–303
30. Weaver JF, Devarajan SP, Hakanoglu C (2009) J Phys Chem C 113: 9773–9782
31. Christy AG, Clark SM (1995) Phys Rev B 52: 9259–9265
32. Huang YS, Liau PC (1987) Chin J Phys 25: 232–242
33. González-Calbet JM, Herrero MP, Alario-Franco MA, Pernet M (1987) J Less-Common Met 135: 105–111
34. Momma K, Izumi F (2011) J Appl Crystallogr 44: 1272–1276
35. Rogal J, Reuter K, Scheffler M (2004) Phys Rev B 69: 075421
36. West AR (2013) Solid State Chemistry and Its Applications. John Wiley & Sons, Ltd., Chichester
37. Antony A, Asthagiri A, Weaver JF (2013) J Chem Phys 139: 104702
38. Kan HH, Weaver JF (2009) Surf Sci 603: 2671–2682
39. He YB, Stierle A, Li WX, Farkas A, Kasper N, Over H (2008) J Phys Chem C 112: 11946–11953
40. Chung WH, Tsai DS, Fan LJ, Yang YW, Huang YS (2012) Surf Sci 606: 1965–1971
41. Kim YD, Schwegmann S, Seitsonen AP, Over H (2001) J Phys Chem B 105: 2205–2211

42. Stoerzinger KA, Qiao L, Biegalski MD, Shao-Horn Y (2014) J Phys Chem Lett 5: 1636–1641
43. Rai R, Li T, Liang Z, Kim M, Asthagiri A, Weaver JF (2016) Surf Sci 652: 213–221
44. Knapp M, Seitsonen AP, Kim YD, Over H (2004) J Phys Chem B 108: 14392–14397
45. Over H, Seitsonen AP, Knapp M, Lundgren E, Schmid M, Varga P (2004) Chem Phys Chem 5: 167–174
46. Weaver JF (2013) Chem Rev 113: 4164–4215
47. Kinnunen NM, Hirvi JT, Suvanto M, Pakkanen TA (2011) J Phys Chem C 115: 19197–19202
48. Van den Bossche M, Gronbeck H (2015) J Am Chem Soc 137: 12035–12044
49. Wang CC, Siao SS, Jiang JC (2012) J Phys Chem C 116: 6367–6370
50. Pham TLM, Leggesse EG, Jiang JC (2015) Catal Sci Technol 5: 4064–4071
51. Tsuji Y, Yoshizawa K (2018) J Phys Chem C 122:15359–15381
52. Erlekam U, Paulus UA, Wang Y, Bonzel HP, Jacobi K, Ertl G (2005) Z Phys Chem 219: 891–903
53. Perdew JP, Burke K, Ernzerhof M (1996) Phys Rev Lett 77: 3865–3868
54. Hall C, Perutz RN (1996) Chem Rev 96: 3125–3146
55. Crabtree RH (1993) Angew Chem Int Ed Engl 32: 789–805
56. Hughbanks T, Hoffmann R (1983) J Am Chem Soc 105: 3528–3537
57. Hoffmann R (1988) Solids and Surfaces: A Chemist's View of Bonding in Extended Structures. Wiley-VCH, Weinheim
58. Mulliken RS (1955) J Chem Phys 23: 1833–1840
59. Chen M, Waghmare UV, Friend CM, Kaxiras E (1998) J Chem Phys 109: 6854–6860
60. Hoffmann R (1963) J Chem Phys 39: 1397–1412
61. YAeHMOP home page. https://yaehmop.sourceforge.net
62. Avogadro: an open-source molecular builder and visualization tool. Version 1.2.0 https://avogadro.cc/
63. Hanwell MD, Curtis DE, Lonie DC, Vandermeersch T, Zurek E, Hutchison GR (2012) J Cheminform 4: 17
64. Avery P, Ludowieg H, Autschbach J, Zurek E (2018) J Chem Educ 95: 331–337
65. Albright TA, Burdett J, Whangbo MH (2013) Orbital Interactions in Chemistry, 2nd ed. Wiley, Hoboken
66. Saillard JY, Hoffmann R (1984) J Am Chem Soc 106: 2006–2026
67. Yoshizawa K (2001) J Organomet Chem 635: 100–109
68. Henkelman G, Uberuaga BP, Jónsson H (2000) J Chem Phys 113: 9901–9904
69. Fung V, Tao F, Jiang D (2018) 20: 22909–22914
70. Tsuji Y, Hoffmann R, Miller JS (2016) Polyhedron 103: 141–149
71. Yoshizawa K, Ohta T, Yamabe T (1998) Bull Chem Soc Jpn 71: 1899–1909

Machine Learning Predictions of Adsorption Energies of CH_4-Related Species

Takashi Toyao, Ichigaku Takigawa, and Ken-ichi Shimizu

Abstract Recent developments in data science could greatly impact catalysis research in both industry and academia. Machine learning (ML) can play a central role in this paradigm shift. Here, we describe a simple and efficient ML approach for predicting the adsorption energies of CH_4-related species, namely CH_3, CH_2, CH, C, and H, on Cu-based alloys. The developed ML model with 12 descriptors, which are readily available for the selected elements, is shown to predict adsorption energies obtained by using density functional theory (DFT)-based calculations. The accuracy and simplicity of the developed system suggest that adsorption energies can be readily predicted without time-consuming DFT calculations. This system will eventually allow the prediction of the catalytic performance of solid catalysts.

Keywords Catalysis informatics · Machine learning · Density functional theory calculations · Brønsted–Evans–Polanyi (BEP) relation

T. Toyao (✉) · K. Shimizu
Institute for Catalysis, Hokkaido University, N-21, W-10, Sapporo 001-0021, Japan
e-mail: toyao@cat.hokudai.ac.jp

K. Shimizu
e-mail: kshimizu@cat.hokudai.ac.jp

Elements Strategy Initiative for Catalysts and Batteries, Kyoto University, Katsura 615-8520, Kyoto, Japan

I. Takigawa
RIKEN Center for Advanced Intelligence Project, 1-4-1 Nihonbashi, Chuo-ku, Tokyo 103-0027, Japan
e-mail: ichigaku.takigawa@riken.jp

Institute for Chemical Reaction Design and Discovery (WPI-ICReDD), Hokkaido University Kita, 21 Nishi 10, Kita-ku, Sapporo 001-0021, Hokkaido, Japan

K. Yoshizawa (ed.), *Direct Hydroxylation of Methane*,
https://doi.org/10.1007/978-981-15-6986-9_7

1 Introduction

Much effort has been devoted to the development of new catalysts for more efficient utilization of CH_4 owing to its economic importance and significant interest in both industry and academia. The development of such catalysis remains challenging because the complexity of the surface reactions and the large number of independent parameters make the prediction of catalyst performance a formidable task. Studies have recently applied descriptors (indicators or relations) to solid catalysts to find a relationship between the bulk/surface structure and performance [1]. In principle, the pace of catalyst discovery could be accelerated by using computational screening methods [2]. Recent developments in methodology and computer technology, as well as the establishment of a descriptor-based approach for the analysis of reaction mechanisms and trends across the periodic table, allow fast screening of new catalytic materials, which has already resulted in the computational discoveries of several new materials [3, 4]. However, these computation-based approaches are typically based on first-principles calculations and thus usually have a high computational cost.

In recent years, machine learning (ML) methods have gained increasing popularity in the molecular and materials science communities for use in the high-throughput screening or the prediction of various kinds of physical properties whose mathematical modeling involves complex principles [5, 6]. ML methods could serve as a fast and high-precision alternative to first-principles modeling. In the ML framework, predictive computations are modeled as a function from some inputs to the output of desired values. Because the input data usually come from experiments, first-principles calculations, or both, applications of ML methods are still limited. Nevertheless, once input data are collected and the proper ML framework is devised, ML methods would enable the fast prediction of various kinds of physical properties.

The automation of the process of intuition in materials synthesis (i.e., from idea to the desired material) is the next step. However, ML methods have just started to be developed for catalysis due to the difficulty of finding a relationship between the bulk/surface structure and catalytic performance [7–9]. Although important contributions have recently been made using ML, the synthesis of truly new catalysts using ML is not currently possible and many aspects of the process remain unexplored. We have recently described the ML prediction of d-band centers [10]. Even though there is no theory-based relationship for sufficiently describing heterogeneous catalysis, d-band centers are widely used as a general and versatile descriptor for various catalytic reactions [11, 12]. However, the accuracy of *d*-band center model predictions for a given catalytic reaction is relatively low. To develop a more accurate catalyst design guide and eventually novel catalysts for challenging reactions such as those that utilize CH_4, more accurate and specific relationships (descriptors) that describe catalytic reactions are required.

To meet this challenge, we have followed ML approaches for predicting adsorption energies. Adsorption is a fundamental step in surface-catalyzed reactions because of the relation between the heat of adsorption and the activation energy, which follows the Brønsted–Evans–Polanyi (BEP) relationship [13].

$$\delta E_{\mathrm{act}} = \mathrm{a}\delta E_{\mathrm{r}} \tag{1}$$

The BEP equation (Eq. 1) directly relates the change in the activation energy of the reaction, δE_{act}, to the corresponding change in the reaction energy, δE_{r}, for different surfaces via a constant factor α, which depends on the particular reaction. A valid BEP relationship enables the prediction of the rate constants for a given family of reactions, as well as related reactions, without the need to perform laborious transition-state searches to determine activation barriers for each reaction. Furthermore, a Sabatier-type trade-off exists between the adsorption of reactants and the desorption of products; catalysts with optimal adsorption energies have the maximal catalytic activity.

Here, we show a protocol of ML prediction for the adsorption energies of CH_4-related species, namely CH_3, CH_2, CH, C, and H, on Cu-based alloys. Using our ML model with 12 descriptors, which are readily available from the periodic table, handbooks, and public databases, we predict adsorption energies calculated using density functional theory (DFT). In addition to predicting the adsorption energies of a single adsorbate, we show that the difference between the adsorption energies of CH_3 and CH_2 on a Cu-alloy surface can also be predicted. Our findings are applicable to the design of selective catalytic reaction processes that suppress undesired reactions, such as the hydroxylation of CH_4, and contribute to the establishment of catalysis informatics, which will enable ML-guided catalyst development.

2 Machine Learning Prediction of Adsorption Energies

2.1 DFT Calculations of Adsorption Energies

DFT-calculated adsorption energies of CH_4-related species, namely CH_3, CH_2, CH, C, and H, on Cu-based alloys were obtained using the Vienna ab initio simulation package (VASP) with projector-augmented wave potentials and the Perdew–Burke–Ernzerhof (PBE) functional [14]. The metal alloys considered in this study have attracted much attention for CH_4 utilization processes [15, 16]. Cu-based alloys were chosen because of their current use as industrial catalysts, economic importance, and application in various CH_4 utilization processes. In our model of Cu-based alloys, the Cu atom located at the center of the surface layer in the slab model is exchanged with another element. For the dopant, we considered elements with atomic numbers of 3 (Li) to 83 (Bi), except for the noble gasses. For ML predictions, models that correctly converged in the DFT calculations and whose optimized structures were not excessively distorted were used. Note that we did not consider whether the alloy structures can exist in reality, because our main objective here is to explore many elements and understand their properties based on ML predictions. However, a recent experimental study on Pt/Cu single-atom alloys suggested that this type of catalyst could be synthesized experimentally and applied to catalytic reactions [17]. CH_3,

CH_2, CH, C, and H species were placed at hexagonal close-packed (hcp)-t, face-centered cubic (fcc)-t, fcc, fcc, and fcc sites, respectively. The initial geometries for each model are shown in Fig. 1.

The obtained values are summarized in Fig. 2. The adsorbates tend to strongly adsorb on Cu-based alloys containing group 5–7 metals. In contrast, the adsorbates are not strongly adsorbed on Cu-based alloys containing group 13–15 metals. These results suggest that the periodic table group is correlated to some extent with the DFT-calculated adsorption energies.

2.2 ML Methods

The DFT-calculated adsorption energies of CH_4-related species were predicted using ML methods. For selecting metal descriptors, we pretested several candidates and chose 12 physical properties for the elements, all of which are readily available from the periodic table, handbooks, and public databases [18, 19]. It should be noted here that for the surface energy descriptor, the value is that for the most stable surface for each metal. To avoid time-consuming DFT calculations while retaining prediction accuracy, the characteristic values were used for descriptors. Each metal was represented as a 12-dimensional vector of the descriptor values. The interdependencies of the descriptors are reflected by the correlation map in Fig. 3, which shows the correlated variables of descriptors. Variable selection was conducted to identify a smaller non-redundant subset of 12 descriptors.

To accomplish data-driven prediction of the adsorption energies, we first separated 46 targets into two disjoint sets, namely a test set of size n and a training set of size

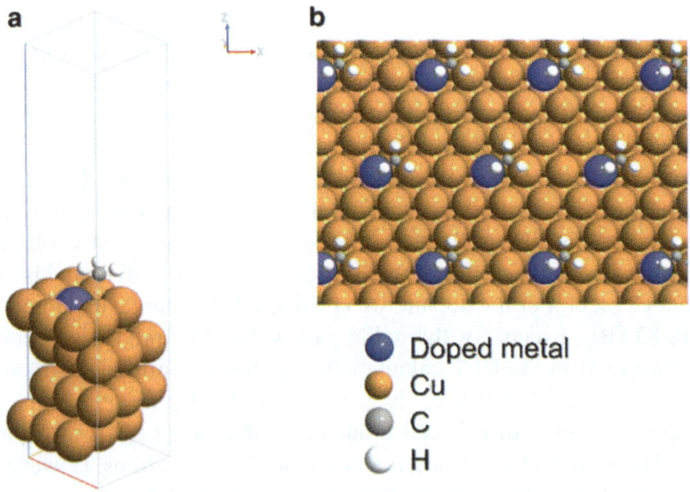

Fig. 1 An adsorption model for CH_3 on Cu-based alloys. **a** Diagonal view and **b** top view

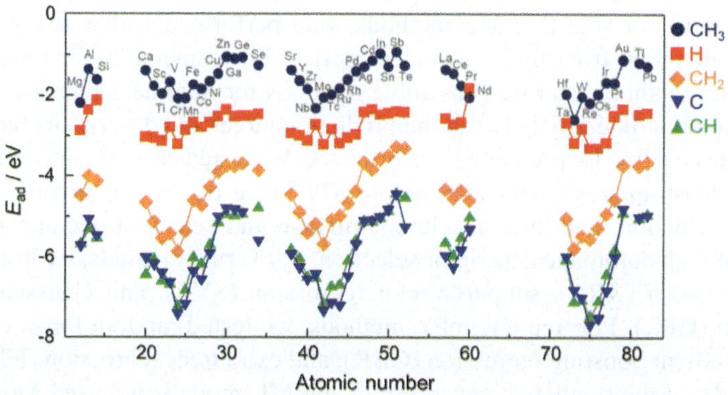

Fig. 2 DFT-calculated adsorption energies of CH_3, CH_2, CH, C, and H on Cu-based alloys. Adapted with permission from Ref. [14]. Copyright 2018 American Chemical Society

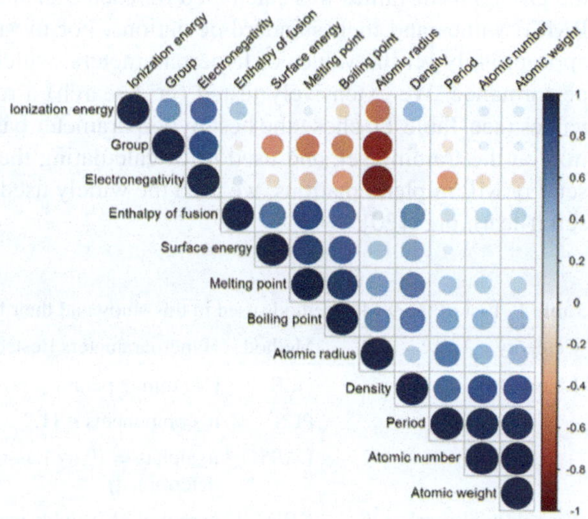

Fig. 3 Correlation maps corresponding to the 12 descriptors used in this study. Adapted with permission from Ref. [14]. Copyright 2018 American Chemical Society

46-*n*. Then, the problem was to evaluate how accurately the adsorption energies of the test set can be predicted using those of the training set. First, an ML model was built using the training set. Then, using this model, the adsorption energies of the test set were predicted, and the root-mean-square error (RMSE) between the predicted values and true values (ground truth) was calculated for prediction evaluation. A single-shot trial of this procedure gives an estimate of RMSE. However, if we change the split of the training and test sets, the estimate would vary with a certain level of variance. For quantitative evaluation, we reduced this estimation variance by repeating the single-shot trial over 100 random test/training splits, i.e., 100 random leave-*n*-out trials, and used the mean of 100 RMSE estimates as the prediction accuracy of the ML model.

Evaluations for selecting ML methods were performed with a set of 9 widely used ML methods from three major categories: linear methods for linear regression, kernel methods, and tree ensemble methods for nonlinear regression. Linear methods assume linearity between the prediction target and descriptors but give the most stable baseline for prediction performance. We tested ordinary linear regression (OLR) by least squares, partial least squares (PLS) regression with automatic dimensionality reduction, and least absolute shrinkage and selection operator (LASSO) regression with automatic descriptor selection. For kernel methods, we tested kernel ridge regression (KRR), support vector regression (SVR), and Gaussian process regression (GPR). For tree ensemble methods, we tested random forest regression (RFR), gradient boosting regression (GBR), and extra trees regression (ETR).

To evaluate the predictive capability of the ML models, we used Monte Carlo cross-validation with 100 random leave-25%-out trials for evaluating the prediction performance for the adsorption energies of CH_3. All ML methods used the 12 descriptors as input. The RMSE of the difference between the predicted values and the ground truth values was calculated for each trial and averaged to obtain the mean RMSE values and their standard deviations. For most ML methods, users need to appropriately set the values of hyperparameters, which critically impact prediction performance. We extensively tested (grid search) a reasonable range of candidate values (see Table 1), chose the best hyperparameter based on threefold cross validation on the training set, and used it for calculating the predicted values for the test set. For ML implementations, we used the widely used package scikit-learn (https://scikit-learn.org) [20].

Table 1 List of the 9 ML methods used in this study and their hyperparameters

Category	Method	Hyperparameters [tested range]
Linear	OLR	(No tuning parameters)
	PLS	n_components ∈ [1,2,…,# of vars]
	LASSO	n_alphas = 10 by LassoCV (Range is automatically determined)
Kernel (nonlinear)	KRR	kernel = 'rbf, alpha, gamma ∈ [1.0,10^{-1},10^{-2},…,10^{-9},10^{-10}]
	SVR	kernel = 'rbf', C ∈ [1.0,10,10^2,…, 10^7,10^8], gamma ∈ [1.0,10^{-1},10^{-2},…,10^{-9},10^{-10}]
	GPR	kernel = Const(1.0, (1e-5, 1e5)) * RBF(1.0, (1e-5, 1e5)) + WhiteKernel(1.0, (1e-5, 1e5)), alpha = 0.0
Tree ensemble (nonlinear)	RFR	n_estimators = 200, max_depth ∈ [1, 2, 3, 4, 8, 10]
	GBR	n_estimators = 200, max_depth ∈ [1, 2, 3, 4, 8, 10] learning_rate ∈ [1.0,10^{-1},10^{-2},…,10^{-5}]
	ETR	n_estimators = 200, max_depth ∈ [1, 2, 3, 4, 8, 10]

Note The default values of scikit-learn were used for the hyperparameters not indicated here

2.3 ML Prediction of Adsorption Energies

Figure 4 shows that the 9 ML methods tested in this study could well predict the adsorption energies of CH_3 within an RMSE of 0.24–0.27 eV on average. This suggests that the adsorption energies of CH_3 on 46 Cu-based alloys share some trends that can be captured from other systems using a data-driven approach. Although the performances of the methods are similar, the differences are statistically significant for ETR versus GPR ($p = 0.04$), ETR versus SVR ($p = 0.02$), ETR versus KRR ($p = 0.01$), and ETR versus OLR ($p = 0.002$) based on the p-values obtained from a Welch two-sample t-test with 100 trials. This suggests that tree ensemble methods are significantly better than kernel methods. We thus used these methods (RFR, GBR, and ETR) for the subsequent analyses in addition to the linear baseline (OLR).

Because the tree ensemble methods are relatively insensitive to hyperparameters, for all subsequent analyses, we used 200 trees in the final ensemble models for RFR, GBR, and ETR. Figure 5 shows the predictive performance for the test set (a random 25%) of the four ML methods fitted to the training set (the remaining 75%). The fourfold cross-validation enables to generate this plot for showing the out-of-sample performance of the ML models because each point in the data belongs to one of the four 25% subparts, and thus must belong to a test set one time in the four 75%/25% splits of the (fourfold) cross-validation procedure. The X-axis represents the DFT-calculated adsorption energies (ground truth) and the Y-axis represents the values predicted by the ML methods. The deviation from the $X = Y$ line indicates the prediction error. For a more quantitative evaluation, we performed 100 random single-shot leave-25%-out trials. The mean RMSE values ($\pm$ standard deviation) for the test set over 100 trials were 0.27 ± 0.07 for OLR, 0.24 ± 0.06 for RFR, 0.24 ± 0.05 for GBR, and 0.24 ± 0.06 eV for ETR (Table 2). The three nonlinear methods (RFR, GBR, and ETR) showed better prediction performance than that of the OLR method ($p < 0.01$ based on Welch two-sample t-test). The three methods had almost identical prediction accuracies. As previously mentioned, the number of regression

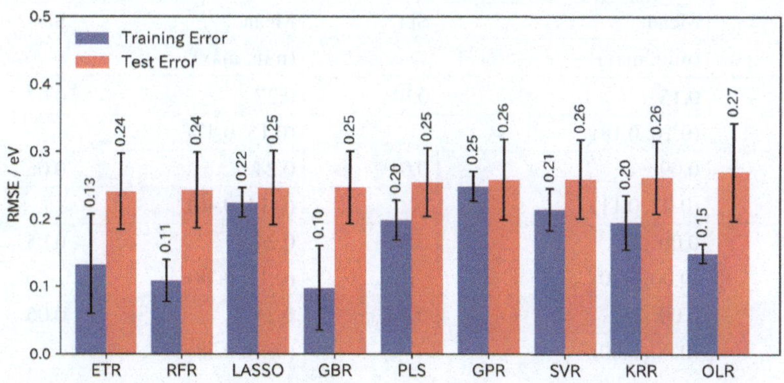

Fig. 4 Average RMSEs for predicting the adsorption energies of CH_3 from 100 random leave-25%-out trials with various ML methods

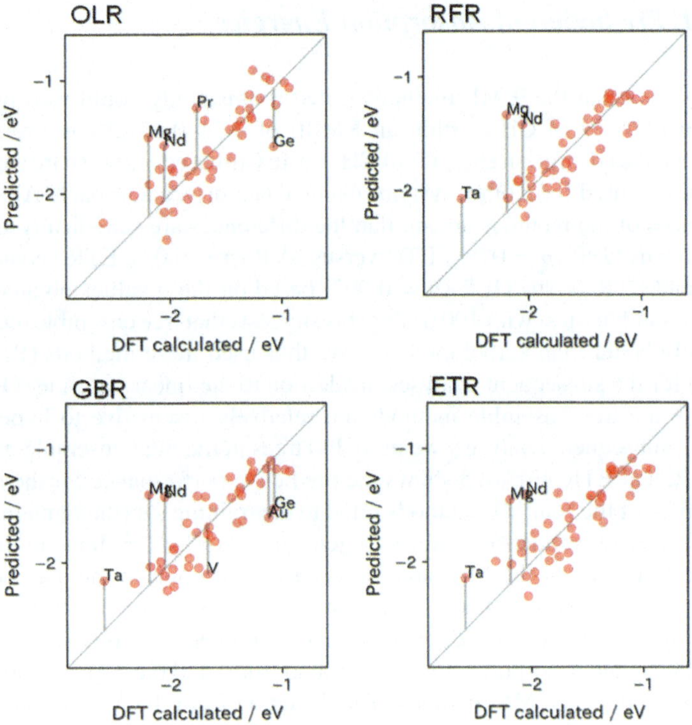

Fig. 5 The out-of-sample prediction performance by fourfold cross-validation; DFT-calculated adsorption energies of CH_3 on Cu-based alloys and values predicted using the OLR, RFR, GBR, and ETR methods. The systems having a large prediction error are indicated in order to provide information on which system is difficult to predict from other systems

Table 2 Mean RMSEs for each ML method for prediction of adsorption energies of CH_3 with respect to values calculated using DFT

	Training error (eV)		Test error (eV)	
	Mean	SD	Mean	SD
	(min, max)		(min, max)	
OLR	0.15	0.01	0.27	0.07
	(0.10, 0.18)		(0.15, 0.47)	
RFR	0.09	0.01	0.24	0.06
	(0.07, 0.11)		(0.14, 0.40)	
GBR	0.00	0.00	0.24	0.05
	(0.00, 0.00)		(0.13, 0.38)	
ETR	0.00	0.00	0.24	0.06
	(0.00, 0.00)		(0.13, 0.38)	

trees used in RFR, GBR, and ETR was fixed at 200; other hyperparameters were set to the default values given by scikit-learn. ETR and RFR are less sensitive to hyperparameters than is GBR, and thus do not require demanding hyperparameter tuning. Moreover, the training time for ETR is shorter than those for RFR and GBR because ETR is based on random-splitting decision trees. Adsorption energies can thus be readily predicted using the ETR method without DFT calculations or hyper-parameter tuning. We therefore conclude that the ETR model is the best choice for the prediction of the adsorption energies considered in this study. The computation times for the DFT calculations and ML analysis were compared. A DFT calculation of the adsorption energy of CH_3 on a Cu monometallic surface took about 10 h on our 32-core workstation, whereas each corresponding ML prediction took less than 1 s on our 1-core laptop computer. The computation time for ML prediction depends on the applied ML model, not the system of interest. For many ML methods that include tree ensembles, the predicted values can be calculated instantly because the function form for prediction is explicitly obtained by fitting an ML model function to the given training data. For prediction, the corresponding descriptor values can be simply substituted into the existing function. This demonstrates the potential of ML as a fast alternative to DFT calculations.

For a better understanding of the ML analysis, we now briefly discuss the prediction results. As shown in our pre-evaluations (Fig. 4), ETR and RFR tend to slightly outperform GBR. Our problem consists of only 46 system examples, some of which are unique, and thus the difficulty of prediction greatly depends on the training and test split. If the training and test sets contain similar systems, the adsorption energies can easily be predicted. In contrast, if the sets contain dissimilar systems, they are difficult to predict. In this situation, it is more important to reduce prediction variance than prediction bias. Both ETR and RFR fit this purpose, with ETR outperforming RFR because it is better at reducing prediction variance. Our current ML model gives some errors even with ETR. This is most likely due to the small dataset size used in this study. For more accurate prediction, it is generally desirable to collect larger datasets.

Next, we show the relevance (or redundancy) of each of the 12 descriptors used in the ETR model. ETR models take a form of an additive ensemble of simple regression-tree models that recursively partition the data points into two using a single selected descriptor for each split. It provides a feature importance score for each descriptor computed as the normalized total reduction of the *impurity* for the generated partitions of the data by that descriptor. This impurity-based score can be used as a standard tool to assess the relative importance of a descriptor with respect to the predictability of adsorption energies. Since this score obtained from the training set can be sometimes misleading, another more careful choice would be the permutation importance score, a decrease in the prediction accuracy for a given trained model when a single feature value in the *test set* is randomly shuffled. Note that although the statistical importance of descriptors varies with the ML method used, these results are only for ETR. Figure 6 shows the feature importance scores and permutation importance scores (averaged with fourfold cross validation) of all 12 descriptors for predicting the adsorption energies of CH_3. The most important

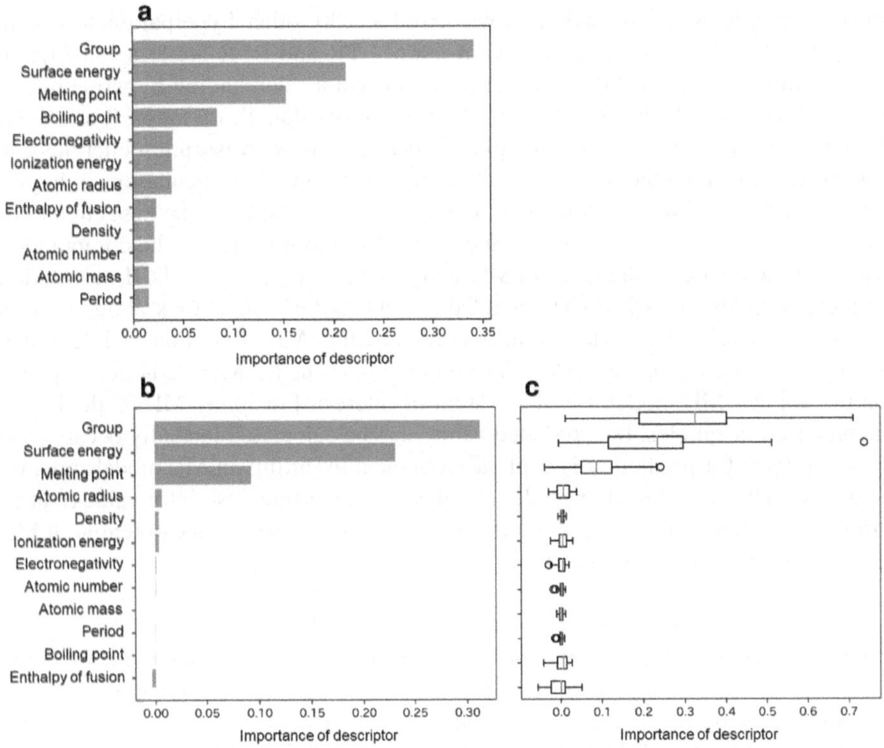

Fig. 6 **a** Feature importance scores and **b**, **c** permutation importance scores of the descriptors for the ETR prediction of the adsorption energies of CH_3 on Cu-based alloys. For permutation importance (**c**), multiple values of the importance scores obtained from fourfold cross-validation with 10 random shuffle for each are visualized using the box plot (the red line shows the median, the box and whisker corresponds the lower/upper quantiles, and minimum/maximum after the outliers indicated as circles are excluded)

descriptor for this system is the periodic table group of the doped metal, followed by surface energy and melting point. It is noteworthy that these descriptors are top ranked for both impurity-based importance scores and permutation importance scores. In addition, one-way and two-way partial dependence plots (PDPs) were computed for the top three descriptors (group, surface energy, and melting point) in order to visualize the marginal effect of the descriptor values on the predicted target response (E_{ads}) by ETR, as shown in Fig. 7. Results clearly show that E_{ads} increases with increase in surface energy and melting point of the doped metal and decrease in periodic table group.

The adsorption energies of CH_2, CH, C, and H were also predicted using the ETR method. Figure 8 shows the results. For the predictions, 100 random single-shot leave-25%-out trials were carried out for each adsorbate to validate the quantitative evaluation. The RMSE values are within 0.2–0.3 eV for all tested adsorbates, suggesting that ML with the ETR method can predict the adsorption energies for

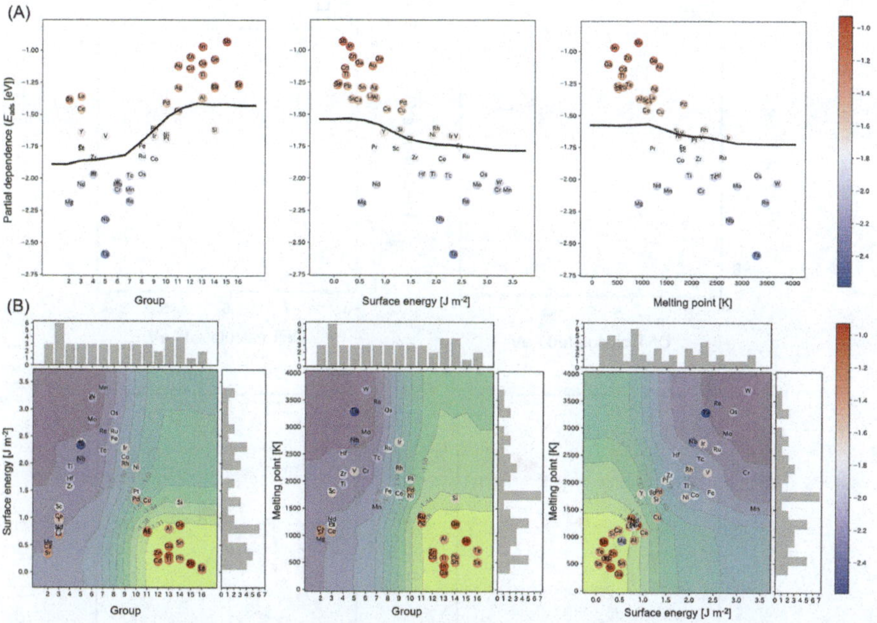

Fig. 7 (**A**) one-way (univariate) and (**B**) two-way (bivariate) partial dependence plots (PDPs) showing the marginal effect of the top three descriptors on E_{ads} of CH_3 for the ETR model. The lines in (**A**) and contours in (**B**) correspond to PDPs, while the histogram and scatter plots of the actual values are also shown in the same figure

various adsorbates without parameter tuning. The feature importance scores of all 12 descriptors for predicting the adsorption energies of CH_2, CH, C, and H were obtained. Interestingly, the periodic table group, surface energy, and melting point are the top three descriptors for all adsorbates, as was the case for CH_3. Moreover, the scores show that prediction ability largely depends on these three descriptors.

2.4 *ML Prediction of* E_{CH3}–E_{CH2} *Values for Methane Utilization*

It is believed that the CH_3 species is a key intermediate for the partial oxidation reaction of methane to methanol. The formation of the CH_3 species is also key for the oxidative coupling of methane (OCM) reaction. These facts suggest that it is important for both methanol synthesis and OCM via the partial oxidation of methane to stabilize the CH_3 species without further dehydrogenation to CH_2, CH, and C, which would eventually yield undesired coke or CO_x formation. For catalyst design, therefore, creating surfaces on which CH_3 species more strongly adsorb than do CH_2

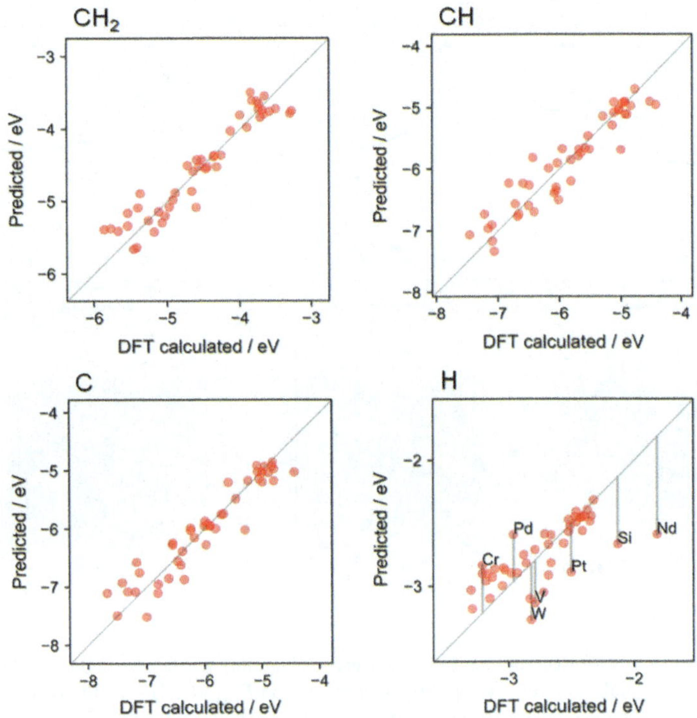

Fig. 8 DFT-calculated adsorption energies of CH_2, CH, C, and H on Cu-based alloys correlated with the values predicted by ETR with 12 descriptors (a four-cross-validation plot showing the out-of-sample prediction performance as the same as that used in Fig. 5). The systems having a large prediction error are indicated in order to provide information on which system is difficult to predict from other systems

species is highly desirable for the efficient utilization of CH_4. We thus compared the adsorption energies of CH_3 and CH_2 on Cu-based alloy surfaces. The difference between the adsorption energies of CH_3 and CH_2 (E_{CH3}–E_{CH2}) was obtained. The values are given in Fig. 9. If our hypothesis is correct, elements that show small E_{CH3}–E_{CH2} values, such as Te, Sn, and Mg, are good candidates for incorporation on Cu surfaces. In contrast, elements that show large E_{CH3}–E_{CH2} values, such as Cr, V, and Mo, could produce surfaces that induce unselective reactions of methane. The obtained E_{CH3}–E_{CH2} values were predicted using four ML methods. Figure 10 shows the predictive performance obtained from 100 random single-shot leave-25%-out trials. The mean RMSE values are also given for each method. Elements for which the obtained error is higher than the standard deviation are shown. The mean RMSE values for OLS, RFR, GBR, and ETR were determined to be 0.35, 0.27, 0.29, and 0.26 eV, respectively. As was the case for the prediction of the adsorption energies of CH_3, RFR, GBR, and ETR outperform OLR. Furthermore, it is advantageous that ETR can be used without parameter tuning. Based on the feature importance analysis, the important descriptors are similar to those for the adsorption energies. The

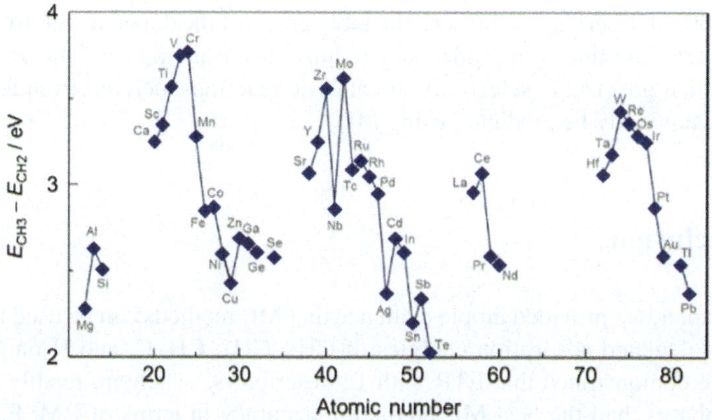

Fig. 9 Difference between DFT-calculated adsorption energies of CH_3 and CH_2 on Cu-based alloys. Adapted with permission from Ref. [14]. Copyright 2018 American Chemical Society

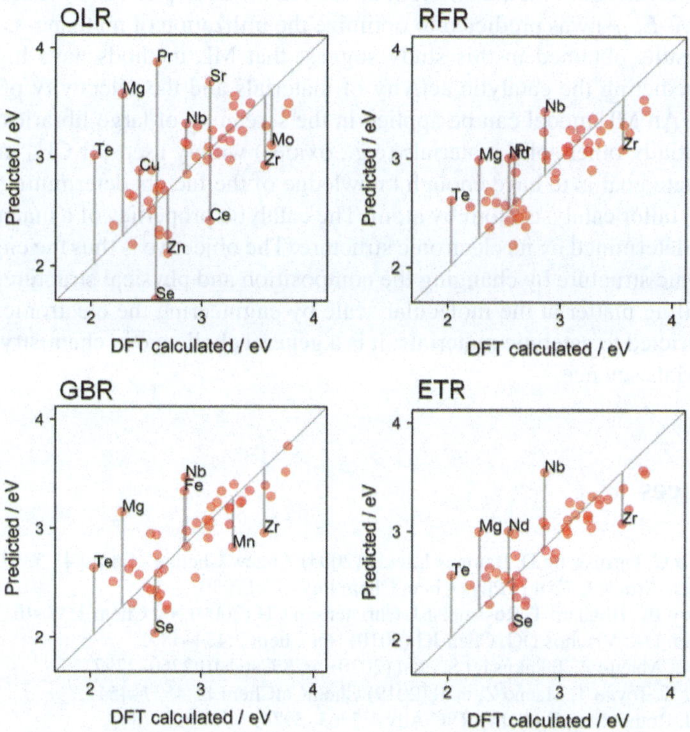

Fig. 10 Difference between DFT-calculated adsorption energies of CH_3 and CH_2 and the values predicted by the OLR, RFR, GBR, and ETR methods (a four-cross-validation plot showing the out-of-sample prediction performance as the same as that used in Figs. 5 and 8). The systems having a large prediction error are indicated in order to provide information on which system is difficult to predict from other systems

most important descriptor is the periodic table group of the doped metal, followed by surface energy, melting point, and atomic radius. This result indicates that E_{CH3}–E_{CH2} values, which govern the selectivity of catalytic reactions, derived from adsorption energies can readily be predicted using ML.

3 Conclusion

In this chapter, we provided ample evidence that ML methods can be used to predict the DFT-calculated adsorption energies of CH_3, CH_2, CH, C, and H on Cu-based alloys. We demonstrated that ETR with 12 descriptors, which are readily available from databases, had the best ML prediction accuracy in terms of RMSE. We also demonstrated that the top three descriptors, namely the periodic table group, surface energy, and melting point, for doped metals can be used to accurately predict adsorption energies. In addition, the difference between the adsorption energies of CH_3 and CH_2 (E_{CH3}–E_{CH2}) was predicted to optimize the utilization of methane.

The results obtained in this study suggest that ML methods have high potential for predicting the catalytic activity of materials and the selectivity of catalytic reactions. An ML model can be applied in the screening of large libraries of alloys and potentially other solid materials (e.g., oxides) widely used for CH_4 utilization. The ultimate goal is to have enough knowledge of the factors determining catalytic activity to tailor catalysts atom by atom. The catalytic properties of a material are in principle determined by its electronic structure. The objective is thus the engineering of electronic structure by changing the composition and physical structure. The aim of controlling matter at the molecular scale by engineering the electronic structure is not restricted to catalytic materials; it is a general challenge in chemistry, physics, and materials science.

References

1. Klanner C, Farrusseng D, Baumes L, et al (2004) Angew Chemie - Int Ed 43:5347–5349.
2. Behler J, Article C (2011) Phys Chem Chem Phys 13:17930.
3. Nørskov JK, Bligaard T, Rossmeisl J, Christensen CH (2009) Nat Chem 1:37–46.
4. Hansgen DA, Vlachos DG, Chen JG (2010) Nat Chem 2:484–489.
5. Toyao T, Maeno Z, Takakusagi S, et al (2020) ACS Catal 10:2260–2297.
6. Suzuki K, Toyao T, Maeno Z, et al (2019) ChemCatChem 11:4537–4547.
7. Ras EJ, Rothenberg G (2014) RSC Adv 4:5963–5974.
8. Coperet C, Comas-Vives A, Larmier K, Copéret C (2017) Chem Commun 53:4296–4303.
9. Bligaard T, Bullock RM, Campbell CT, et al (2016) ACS Catal 6:2590–2602.
10. Takigawa I, Shimizu K, Tsuda K, Takakusagi S (2016) RSC Adv 6:52587–52595.
11. Ruban A, Hammer B, Norskov JK (1997) J Mol Catal A, Chem 115:421–429
12. Nørskov JK, Abild-Pedersen F, Studt F, Bligaard T (2011) Proc Natl Acad Sci U S A 108:937–943.
13. Santen RAV, Neurock M, Shetty SG (2010) Chem Rev 110:2005–2048.

14. Toyao T, Suzuki K, Kikuchi S, et al (2018) J Phys Chem C 122:8315–8326.
15. Besenbacher F, Chorkendorff I, Clausen BS, et al (1998) Science 279:1913–1915.
16. Kokalj A, Bonini N, Gironcoli S De, et al (2006) J Am Chem Soc 128:12448–12454
17. Marcinkowski MD, Darby MT, Liu J, et al (2018) Nat Chem 10:325–332.
18. Lide DR (2003) CRC Handbook of Chemistry and Physics, 84th Edition, 2003–2004. Handb Chem Phys 53:2616.
19. Tran R, Xu Z, Radhakrishnan B, et al (2016) Surface energies of elemental crystals. Sci Data 3:160080.
20. Pedregosa F, Varoquaux G (2011) Scikit-learn: Machine learning in Python. J Mach Learn Res 12:2825–2830.

Theoretical Approach to Homogeneous Catalyst of Methane Hydroxylation: Collaboration with Computation and Experiment

Yuta Hori and Tsukasa Abe

Abstract In the past decades, quantum chemical calculations such as density functional theory (DFT) have played important roles in catalytic reactions, which provide important guidelines and practical strategies for rational catalytic design. Because direct conversion of methane to methanol is a major challenge in catalysis, theoretical approach requires for the catalyst design in methane hydroxylation. In the present review, examples are given of applications of DFT to the reaction mechanisms and successful predictions of methane hydroxylation catalyzed by homogeneous catalysts and developed by the interplay with theory and experiment. Platinum complexes as organometallic catalysts and diiron, dicopper, and tricopper complexes as biomimetic catalysts are introduced.

Keywords DFT calculations · Homogeneous catalyst · Methane hydroxylation · Transition metal complex

1 Introduction

Methane is an abundant and inexpensive gas, the major component in natural gas, and a promising resource for energy and chemical production [1–3]. The utilization of natural gas is limited by the difficulty of transport because long-distance transport of gas-formed methane through pipelines is impractical. The transport of the liquefaction of natural gas is also not a viable option because of harsh conditions (the boiling point of methane is 109 K at 1 atm). An efficient use of methane in natural gas requires its chemical conversion into a liquid product such as methanol. Methanol is also an efficient and flexible feedstock for the production of chemicals

Y. Hori (✉) · T. Abe
Institute for Materials Chemistry and Engineering and IRCCS, Kyushu University, Fukuoka 819-0395, Japan
e-mail: hori@ms.ifoc.kyushu-u.ac.jp

T. Abe
e-mail: abe@ms.ifoc.kyushu-u.ac.jp

K. Yoshizawa (ed.), *Direct Hydroxylation of Methane*,
https://doi.org/10.1007/978-981-15-6986-9_8

such as dimethyl ether, formaldehyde, and propylene, in place of petroleum. Therefore, chemical conversion from methane to valuable chemicals has captivated the interest of not only the academic community for more than a century but also of industry.

The conversion of methane into methanol has been implemented in industry as follows: (i) methane is first converted into synthesis gas (a mixture of carbon monoxide and hydrogen) by steam reforming [Eq. (1)], and (ii) the synthesis gas is then converted into methanol [Eq. (2)].

$$CH_4 + H_2O \rightarrow CO + 3H_2 \quad \Delta H^0_{298\,K} = +206.2\ \text{kJ mol}^{-1} \tag{1}$$

$$CO + 2H_2 \rightarrow CH_3OH \quad \Delta H^0_{298K} = -90.7\ \text{kJ mol}^{-1} \tag{2}$$

Steam reforming, catalyzed by alumina-supported nickel, proceeds at temperatures as high as 1100 K and results in significant operating and maintenance costs. On the other hand, the direct hydroxylation of methane is more energy efficient [Eq. (3)] than the indirect process via steam reforming, because methane hydroxylation is an exothermic process [4].

$$CH_4 + 0.5O_2 \rightarrow CH_3OH \quad \Delta H^0_{298\,K} = -126.2\ \text{kJ mol}^{-1} \tag{3}$$

Therefore, efficient catalyst design for the partial oxidation of methane to methanol has been one of the most important goals of industry.

Direct conversion of methane to methanol is a major challenge in catalysis because there are two hurdles to overcome in the oxidation process—the difficulty in activating the C–H bond of methane and the challenge thermodynamically to preserve the desired methanol product without producing overoxidized products, such as CO_2 or coke. Methane is a stable and inert hydrocarbon owing to its perfectly symmetrical tetrahedral structure, non-polar nature, and its C–H bond dissociation energy (BDE) of about 104 kcal mol^{-1}. Thus, C–H bond activation may require harsh conditions. In contrast, because the BDE of the C–H bonds of methanol is 11 kcal mol^{-1} lower than that of methane [5], methanol is highly susceptible to overoxidation to CO_2 under the reaction conditions in which the C–H bond of methane is activated. Thus, an effective catalyst would have to advance methane activation (conversion) and simultaneously prevent methanol oxidation (selectivity) [2].

The partial oxidation of methane to methanol has been carried out in four ways, namely high-temperature non-catalytic gas-phase oxidation, heterogeneous catalytic oxidation, low-temperature homogeneous catalytic oxidation in solution, and enzymatic or biological catalytic oxidation [1]. In this chapter, we focus on homogeneous catalysis in solution, based on the organometallic and biomimetic approaches. An advantage of methane hydroxylation using homogeneous catalysts is that methanol is produced more selectively by homogeneous catalysts than by heterogeneous catalysts operating at high temperatures, because in the systems using homogeneous catalysts, methane activation is performed at relatively low temperatures leading to suppress the overoxidation reaction kinetically.

Theoretical approaches using computational chemistry play important roles in catalytic reactions, which provide important guidelines and practical strategies for rational catalytic design [6, 7]. Theoretical calculations often assist in the interpretation of experimental data, such as spectroscopic data, and provide mechanistic insight into catalytic reactions. In addition, theoretical calculations sometimes give us preliminary information of catalytic performance that is subsequently tested experimentally. As developments in hardware and software lead to faster computation, the theoretical approach to catalyst design of methane hydroxylation might eventually become important. Computational chemistry has been employed not only by theoreticians but also by experimental groups to support experimental data. The direct interaction between the theoretical results and experimental data helps the development of new and improved methane hydroxylation catalysts. The feedback between theory and experiment is especially critical for success of the catalytic design of methane hydroxylation [6].

In this chapter, we introduce some catalytic complexes developed by the interplay with theory and experiment and several examples of successful predictions for methane hydroxylation using computational chemistry.

2 Computational Methods

A wide range of theoretical methods is available for catalyst design in methane hydroxylation. For molecular systems, in particular, quantum chemical calculations using methods such as density functional theory (DFT), which is the most practical method for reactions involving transition metal complexes, are able to predict reliable geometrical and electronic structures, thermodynamics properties, and reaction pathways for specific reactions. A conventional approach to understanding the reaction mechanisms of catalytic system by quantum chemical calculations starts from presuming the transition state (TS) and the reaction pathway for the each step of desired reaction followed by optimization of the stationary points, such as reactant complex (RC), TS, and product complex (PC) and calculating the intrinsic reaction coordinate (IRC). Performing these tasks in all steps in the catalytic reactions, we can find a reasonable single reaction path connected the overall reaction. These results provide the relative energies of the RC, TS, and PC in the each reaction and the activation energy and reaction energy in the catalytic reaction. In quantum chemical calculations of the transition metal complexes, in particular, the electronic configurations and spin states of the desired complexes need to be considered [7]. Because the energy levels of various spin states in the transition metal complexes, such as the high spin or low spin state, might be in a range of a few kcal mol^{-1}, the energies and spectra of different spin states should be computed and compared with either experimental data or a higher-level theory. In addition, in the metal-oxo and polynuclear metal complexes, there might be the open-shell or closed-shell states in the same spin state and the ferromagnetic coupled or antiferromagnetic coupled state. A broken-symmetry approach using DFT can treat different electronic configurations, although spin contamination becomes a problem. The solvent, which plays an important role in catalytic reactions, can be accounted for by using an implicit solvent model (where the solvent is treated as a dielectric continuum) or an explicit solvent model (where solvent molecules are treated at a lower level of quantum mechanics or using molecular mechanics). Enzymatic systems are often modeled by a mixed quantum mechanical/molecular mechanical (QM/MM) model, in which active sites are treated quantum mechanically, and the remainder of the system is treated molecular mechanically. The choice of computational methods and models depends on the level of accuracy required and the scale of the system [6].

3 Organometallic Approaches

Methane activation at low temperatures has been investigated by using homogeneous catalysis. The C–H bond activation of methane under low temperatures does not generate radical species and may lead to more selective reactions than those promoted by heterogeneous catalysts operating at high temperatures.

Shilov reported some of the earliest evidence that the inert C–H bond of several alkanes, including methane, were activated in a catalytic fashion [8]. Shilov showed that H/D exchange would occur between alkanes and deuterium acid by the Pt(II) and Pt(IV) complexes. Although this reaction is impractical owing to the use of the Pt(IV) complex as the oxidant with high cost, these results gave an indication that methane functionalization could be achieved by organometallic approaches, and the transformation, known as the Shilov reaction, was developed in the period 1970–1985.

Based on the Shilov reaction, Periana reported two different systems for the oxidation of methane in sulfuric acid containing SO_3. One of the catalysts was a simple Hg(II) halide, and the Hg(II)-catalyzed reactions generated methyl sulfate with a turnover frequency (TOF) of $10^{-3}\ s^{-1}$ [9]. The second system was more reactive and was based on a complex of Pt(II) with a bipyrimidine (**bpym**) derivative [10]. In this case, methane was converted to methyl bisulfate (CH_3OSO_3H) [Eq. (4)] with 81% selectivity, >500 turnovers, and a TOF of $10^{-2}\ s^{-1}$. The thus obtained methyl bisulfate was then readily hydrolyzed to produce methanol [Eq. (5)].

$$CH_4 + 2H_2SO_4 \rightarrow CH_3OSO_3H + SO_2 + 2H_2O \tag{4}$$

$$CH_3OSO_3H + H_2O \rightarrow CH_3OH + H_2SO_4 \tag{5}$$

In this system, methane selectively converts to methanol because the electron-withdrawing group attached to the oxygen protects the overoxidation of methanol. However, the major disadvantages are difficulty in separating the methanol product from sulfuric acid used as a solvent system, need for expensive corrosion-resistant materials due to the corrosive nature, and periodic regeneration of the acid [11].

Mechanistic studies have been conducted on both Shilov oxidations and related oxidations reported by Periana to develop an optimal methane hydroxylation catalyst [12]. These oxidations typically start with the C–H bond activation of methane, which results in the formation of a Pt(II) methyl complex from a Pt(II) halide or hydroxide complex. The Pt(II) methyl complex is then oxidized to a Pt(IV) methyl complex by transfer of the halide from the Pt(IV) to the Pt(II) complex. Finally, a carbon–halogen bond-forming step by attack of water, hydroxide, sulfonate, or halide on the resulting Pt(IV) methyl complex generates the methanol or methyl halide product and regenerates the starting Pt(II) complex. The most extensive studies on the detailed mechanism of this process focused on how the C–H bond activation of methane occurs. Several theoretical studies have been performed for the mechanism of the Shilov and Periana reactions, as summarized in the review of Eisenstein and co-workers [13].

Siegbahn and Crabtree theoretically studied the Shilov reaction using DFT calculations [14]. The C–H bond activation of methane by *trans*-[Pt$(H_2O)_2(Cl)_2$] was considered as a model reaction, and the free energies along the reaction pathway for the reaction between [Pt$(H_2O)_2(Cl)_2$] and methane were calculated, the results of which are shown in Fig. 1. The reaction starts from the substitution of one H_2O ligand by CH_4 with an energy cost of about 10 kcal mol^{-1}. The next step proceeds

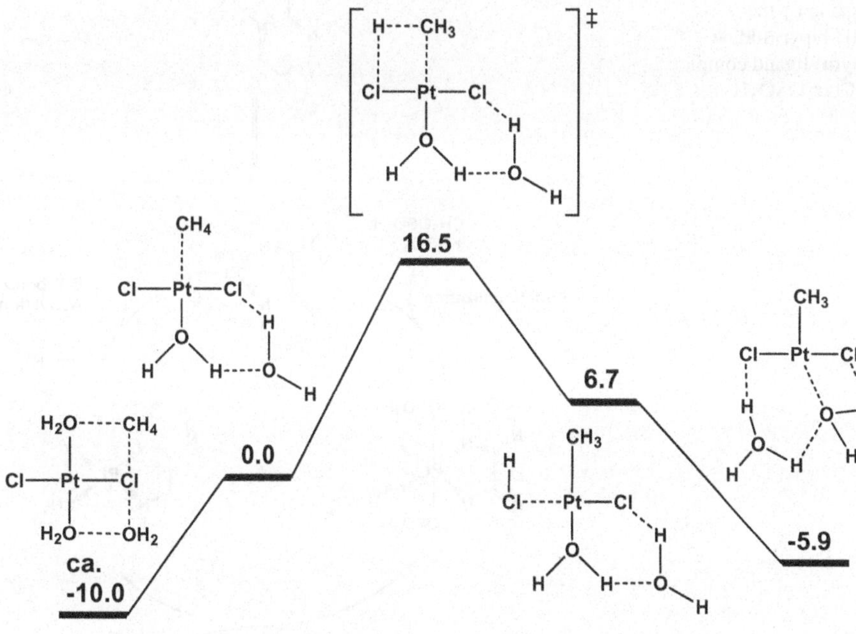

Fig. 1 Free energies calculated along the reaction pathway for the reaction between [Pt $(H_2O)_2Cl_2$] and methane [14]. Relative energies are in kcal mol^{-1}

like σ-bond metathesis with an energy barrier of 16.5 kcal mol^{-1}, in which the transformation of an H atom in CH_4 to a Cl ligand affords HCl and a Pt–CH_3 bond. The overall activation energy including the initial substitution step is about 27 kcal mol^{-1}, in agreement with the experimental value of 28 kcal mol^{-1} [14, 15]. In the final step, intramolecular proton transfer to one of the H_2O molecules, which involves a swing of the HCl hydrogen, yields the $[PtCl_2(CH_3)(H_5O_2)]$ complex with a very small energy barrier, stabilizing the system with 12.6 kcal mol^{-1}.

Periana proposed a reaction mechanism for the catalytic oxidation of methane by the Pt(II) complex with a **bpym** ligand, as shown in Fig. 2. A 14-electron cationic Pt(II) complex first reacts with methane to form a Pt(II) methyl complex. The oxidation of the Pt(II) methyl then leads to a Pt(IV) methyl sulfonate complex. Finally, the reductive elimination of methyl bisulfate and loss of a bisulfate ligand from the Pt center occurs to regenerate the Pt(II) complex.

Gilbert et al. explored the mechanism of methane C–H bond activation by $[Pt(\mathbf{bpym})(Cl)]^+$ and $[Pt(\mathbf{bpym})(OSO_3H)]^+$ using DFT calculations [16]. C–H bond activation was proposed to occur via oxidative addition with an energy barrier of 10.0 kcal mol^{-1} if $[Pt(\mathbf{bpym})Cl]^+$ acts as the active species (Fig. 3a), while σ-bond metathesis was suggested to take place if $[Pt(\mathbf{bpym})(OSO_3H)]^+$ acts as the active

Fig. 2 Reaction mechanism proposed by Periana for the catalytic oxidation of methane by the Pt(II)-bipyrimidine (**bpym**) ligand complex. X=Cl or OSO_3H

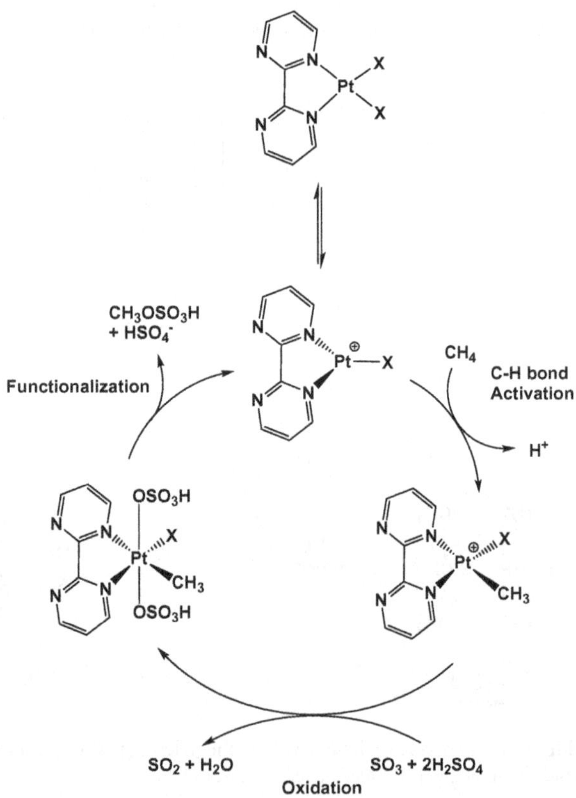

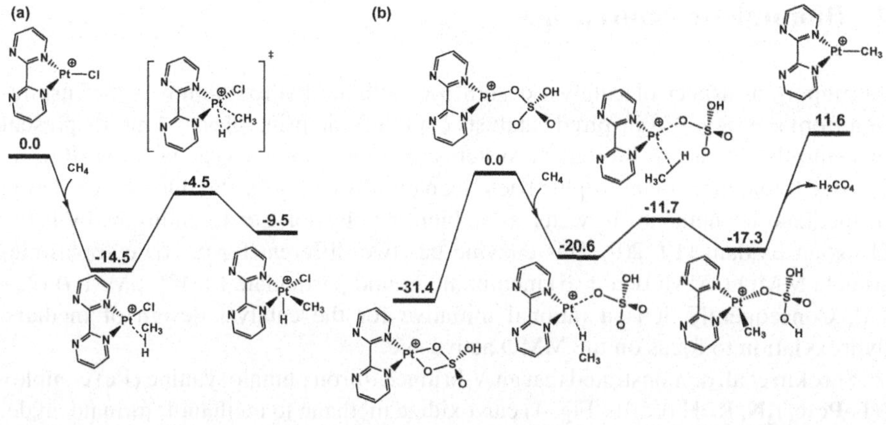

Fig. 3 Calculated energy diagrams of methane activation for **(a)** oxidative addition in the case of [Pt(**bpym**)Cl]$^+$ and **(b)** metathesis in the case of [Pt(**bpym**)(OSO_3H)]$^+$. Relative energies are in kcal mol^{-1}

species (Fig. 3b). Thus, the C–H bond activation mechanism depends on the nature of the ligands.

For the catalyst design of methane hydroxylation, these mechanistic studies using theoretical calculations demonstrate that the overall C–H bond activation process is controlled by two key steps in the Shilov and Periana systems: (1) coordination of methane to Pt center and (2) the C–H bond cleavage of methane. From this knowledge, we can predict that in the design of an improved catalyst for methane hydroxylation, it is important to reduce the energetics for methane coordination, because the Pt complexes have enough potential for the C–H bond cleavage of methane. Goddard and Periana reported that an efficient system has been designed that catalyzes the H/D exchange much faster than the [Pt(**bpym**)Cl_2] system does, by orders of magnitude, using the interplay between computational and experimental methods [17]. They postulated that [Pt(**pic**)(TFA)$_2$]$^-$ (pic$^-$ = η^2-N,O-picolinate, TFA$^-$ = trifluoroacetate) would have reduced energetics for hydrocarbon coordination because of the increased electron density at the metal center. These hypotheses were confirmed by comparing the catalytic properties of [Pt(**bpym**)(TFA)$_2$] and [Pt(**pic**)(TFA)$_2$]$^-$ in the oxidation of benzene with H_2SO_4. The experiments showed that the [Pt(**pic**)(TFA)$_2$]$^-$ catalyst is 300 times more active than [Pt(**bpym**)(TFA)$_2$]. In addition, DFT calculations showed that the coordination of benzene in [Pt(**pic**)(TFA)$_2$]$^-$ involves an energy barrier of 5.0 kcal mol^{-1}, which is almost three times lower than that in [Pt(**bpym**)(TFA)$_2$], in agreement with their hypotheses.

Tsuji et al. investigated the nature of the adsorption and activation of methane on the surface of rutile-type metal dioxides IrO_2, CrO_2, and PtO_2, using first-principle calculations, and suggested that distorted rutile-type dioxide β-PtO_2 shows high methane activation reactivity [18].

4 Biomimetic Approaches

An important aspect of catalyst design for methane hydroxylation is the inspiration from biology. Bio-inspired catalysts exploit basic principles and motifs present in naturally occurring biological systems. Methane monooxygenase (MMO), an enzyme found in methanotrophic bacteria, naturally catalyzes the selective oxidation of methane to methanol in water at ambient or physiological conditions by using O_2 as an oxidant [19, 20]. This enzyme has two different forms: (i) cytoplasmic-soluble MMO (sMMO) and (ii) membrane-bound particulate MMO (pMMO) [21–23]. Consequently, it is a rational initiative for the catalyst design of methane hydroxylation to focus on the MMO active site.

Sorokin et al. demonstrated that an *N*-bridged diiron phthalocyanine (**Pc**) complex (**(FePcR$_4$)$_2$N**, R=H or ^{t}Bu; Fig. 4) can oxidize methane to methanol, formaldehyde, and formic acid in the presence of H_2O_2 in water at ambient temperature (25–60 °C) [24].

Table 1 summarizes the results of methane oxidation tests performed at different temperatures, showing that even at 25 °C, oxidation was efficient and afforded formic acid with a turnover number (TON) of 13 (Table 1, run 1). At 40–50 °C, formic acid and formaldehyde were obtained in a ~2:1 ratio (Table 1, runs 2 and 3). At higher temperatures, the amount of formaldehyde diminished in favor of formic acid. The catalytic activity was similar between 40 and 80 °C, providing 26–32 turnovers. The heterolytic O–O bond cleavage in the $Fe^{IV}NFe^{III}OOH$ complex and the formation of the putative very strongly oxidizing $Fe^{IV}NFe^{V}{=}O$ species should be favored in the presence of acid by the protonation of oxygen peroxide. Indeed, a significant improvement in the catalytic activity was observed in the presence of 0.1 M H_2SO_4, and TON_{HCOOH} increased to 72.8. After the completion of the first reaction, a new portion of H_2O_2 was added directly to the reaction mixture. Remarkably, the catalytic system retained practically the same catalytic activity in the second cycle (Table 1, run 7), indicating high catalytic stability and even a possibility of recycling. The catalyst exhibits very high performance: more than 150 mol of CH_4 per mole of

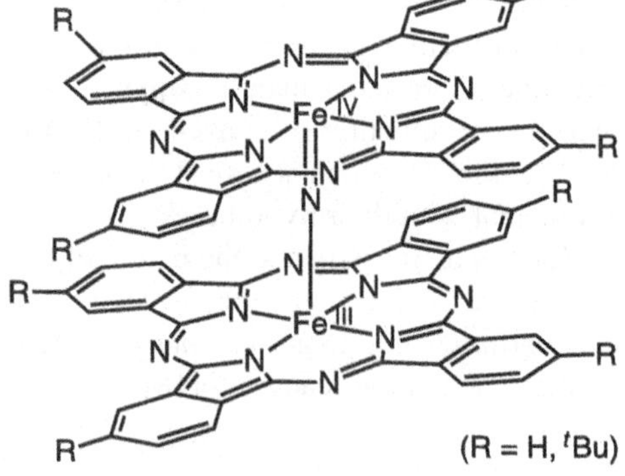

Fig. 4 μ-Nitrido-bridged diiron phthalocyanine (**Pc**) complexes developed by Sorokin for methane hydroxylation. R=H or ^{t}Bu for the μ-nitrido complex of iron phthalocyanine **(FePc)$_2$N** and the iron tetra-*tert*-butylphthalocyanine **(FePc^{t}Bu$_4$)$_2$N**, respectively [24]

Table 1 Oxidation of methane by H_2O_2 in water catalyzed by $\mathbf{(FePc^tBu_4)_2N}$[a]

Run	T (°C)	[HCOOH] (mM)	TON_{HCOOH}	[HCHO] (mM)	TON_{HCHO}	Total TON[b]
1	25	6.0	13.0	0	0	39.0
2	40	8.6	18.6	4.8	10.4	76.6
3	50	9.2	21.0	4.7	10.7	84.4
4	60	10.5	22.8	1.5	3.2	74.8
5	70	11.7	25.2	0.8	1.7	79.0
6	80	12.8	27.3	0.5	1.1	84.1
7[c]	60	69.0 (34.1)	134.6 (72.8)	7.6	16.5	436.8

[a]Conditions: 32 bar CH_4; 2 mL H_2O; catalyst, 0.925 mmol (0.875 mmol for run 3); 678 mmol H_2O_2; reaction time 20 h (48 h for run 1)
[b]Total TON was calculated 3 × HCOOH/catalyst + 2 × $CH_2(OH)_2$/catalyst
[c]In 0.1 M H_2SO_4, 678 mmol H_2O_2 were added at reaction times 0 and 16 h. Values in parentheses were measured before the second addition of H_2O_2

catalyst were oxidized to useful products. This activity is far higher than that of most published systems, operating via methane activation [9, 15, 25–28].

Sorokin proposed the reaction mechanism shown in Fig. 5 [24]. In the first step, the diiron complex $\mathbf{Fe^{IV}NFe^{III}}$ coordinates to H_2O_2 to form the hydroperoxo complex $\mathbf{Fe^{IV}NFe^{III}OOH}$. The heterolytic O–O bond cleavage in the $\mathbf{Fe^{IV}NFe^{III}OOH}$ should be favored to form the putative very strongly oxidizing species $\mathbf{Fe^{IV}NFe^{V}{=}O}$. Then, the $\mathbf{Fe^{IV}NFe^{V}{=}O}$ complex should oxidize methane to give oxygenated products and regenerate $\mathbf{Fe^{IV}NFe^{III}}$, completing the catalytic cycle.

To get further mechanical insight into the reaction, Rajaraman and co-workers conducted DFT calculations for methane hydroxylation by using diiron

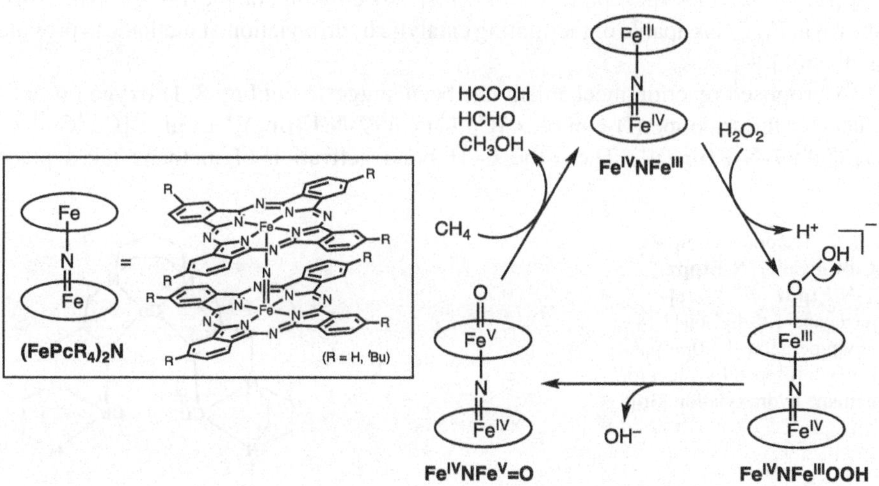

Fig. 5 Proposed mechanism for methane hydroxylation catalyzed by $\mathbf{(FePcR_4)_2N}$

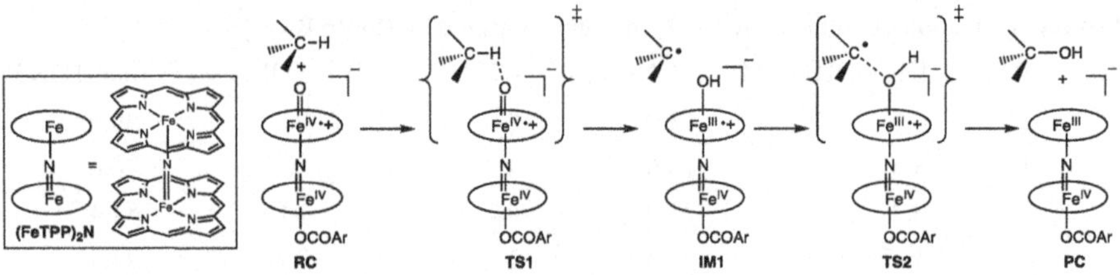

Fig. 6 Proposed mechanism for methane hydroxylation catalyzed by $\mathbf{(FeTPP)_2N}$

complexes supported by an *N*-bridged diiron tetraphenylporphyrine (**TPP**) complex ($\mathbf{(FeTPP)_2N}$) [29]. For the active species, they found that $[Fe^{IV}(\mathbf{TPP})(\mu\text{-}N)Fe^{IV}(\mathbf{TPP}^{\cdot+})(O)]^-$ in the antiferromagnetically coupled doublet state is more favorable than $[Fe^{IV}(\mathbf{TPP})(\mu\text{-}N)Fe^{V}(\mathbf{TPP})(O)]^-$. In the reaction mechanism shown in Fig. 6, the C–H bond activation of methane by the $Fe^{IV}{=}O$ unit (**RC**) is assumed, giving an Fe^{III}–OH intermediate (**IM1**) via **TS1**. Then, rebounding to the OH group is assumed to take place via **TS2**, leading to the formation of methanol as well as the Fe(III) complex (**PC**).

The *μ*-nitrido diiron complex, supported by the phthalocyanine ligand ($\mathbf{(FeTPP)_2N}$), exhibits highly catalytic performance for methane hydroxylation. DFT calculations revealed the reason why the dinuclear system can be an efficient catalyst for methane hydroxylation through a clarification of a subtle electronic structure of a *μ*-nitrido bridged dinuclear Fe^{IV}–oxo species, as well as the reaction mechanism for methane oxidation. The idea of electronic cooperativity as presented here has the potential for wider application in other diion models.

In another example for methane hydroxylation using the biomimetic approach, Chan et al. reported that the tricopper complex $[Cu^ICu^ICu^I(\mathbf{7\text{-}N\text{-}Etppz})]^+$ (**7-N-Etppz** = 3,3′-(1,4-diazepane-1,4-diyl)bis[1-(4-ethylpiperazine-1-yl)propan-2-ol]) shown in Fig. 7 is capable of facilitating catalytic hydroxylation of methane to provide methanol [30].

A proposed reaction mechanism has been suggested in Fig. 8. Dioxygen is activated by the tricopper(I) complex $[Cu^ICu^ICu^I(\mathbf{7\text{-}N\text{-}Etppz})]^+$ to give $[Cu^{II}Cu^{II}(\mu\text{-}O)_2Cu^{III}(\mathbf{7\text{-}N\text{-}Etppz})]^+$. Then, the C–H bond activation of methane takes place

Fig. 7 Tricopper complex $[Cu^ICu^ICu^I(\mathbf{7\text{-}N\text{-}Etppz})]^+$ (**7-N-Etppz** = 3,3′-(1,4-diazepane-1,4-diyl)bis[1-(4-ethylpiperazine-1-yl)propan-2-ol]) developed by Chan for methane hydroxylation [30]

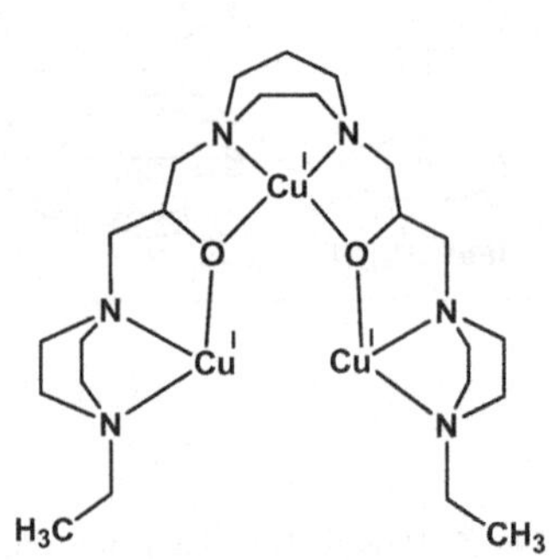

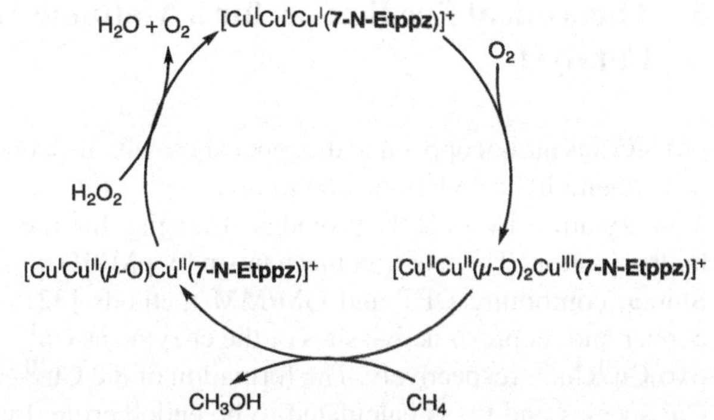

Fig. 8 Proposed reaction mechanism of the hydroxylation of methane catalyzed by tricopper complex $[Cu^ICu^ICu^I$(**7-N-Etppz**)$]^+$

to give methanol as well as $[Cu^ICu^{II}(\mu$-O)Cu^{II}(**7-N-Etppz**)$]^+$. Hydrogen peroxide can regenerate the tricopper(I) complex $[Cu^ICu^ICu^I$(**7-N-Etppz**)$]^+$, completing the catalytic cycle.

To get further insight into the reaction mechanism, Jiang and co-workers conducted DFT calculations for methane hydroxylation by $[Cu^{II}Cu^{II}(\mu$-O$)_2Cu^{III}$(**7-N-Etppz**)$]^+$ [31]. It was found that CH_4 can interact with the activated tricopper complex to form a hydrogen bond between one of the C–H bonds in CH_4 and the O_2 molecule, activating the tricopper cluster complex. In addition, the concerted electrophilic oxene insertion mechanism and the nucleophilic hydrogen-atom abstraction/geminal radical rebound mechanism were considered for the process. In the former mechanism, the C–H bonds in CH_4 occur through the nonlinear C–O–H transition state and are accompanied by direct O-atom insertion to form CH_3OH. Calculated energy diagrams shown in Fig. 9 indicate that the latter process was found to have a smaller activation energy of 8.8 kcal mol^{-1} than the former with an activation energy of 16.1 kcal mol^{-1}.

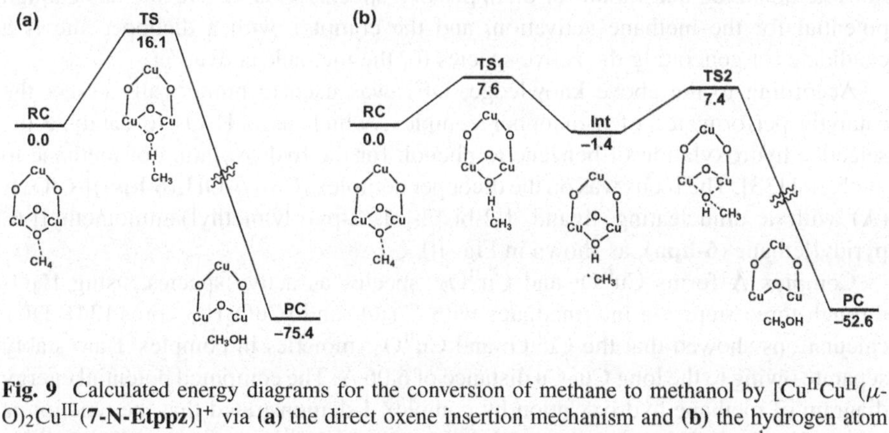

Fig. 9 Calculated energy diagrams for the conversion of methane to methanol by $[Cu^{II}Cu^{II}(\mu$-O$)_2Cu^{III}$(**7-N-Etppz**)$]^+$ via **(a)** the direct oxene insertion mechanism and **(b)** the hydrogen atom abstraction/geminal radical rebound mechanism. Relative energies are in kcal mol^{-1}

5 Theoretical Predictions for a Methane Hydroxylation Catalyst

pMMO has monocopper and dicopper sites, which is reported to play an essential role in methane hydroxylation. The knowledge of the reaction mechanism of methane hydroxylation by pMMO provides strategies for the catalytic design of methane hydroxylation. The reaction mechanism by pMMO was explored by Yoshizawa and Shiota, combining DFT and QM/MM methods [32]. They considered the monocopper and dicopper active sites of the enzyme as Cu^{III}–oxo (or Cu^{II}–O·) and bis(μ-oxo)$Cu^{II}Cu^{III}$, respectively. The formation of the Cu^{III}–oxo species by reaction with Cu^{I} species and O_2 is calculated to be endothermic by 17.8 kcal mol^{-1}, while the formation of the bis(μ-oxo)$Cu^{II}Cu^{III}$ species by reaction with the dicopper site and O_2 is exothermic by 59.3 kcal mol^{-1}. QM/MM calculations show that the Cu^{III}–oxo species is coordinated by His48, His72, and Glu75, and bis(μ-oxo)$Cu^{II}Cu^{III}$ species is coordinated by His33, His137, His139, and Glu35. In the dinuclear copper site, the formal charge of the copper ion coordinated by Glu35 is assigned to be +3 and that of the other copper ion is +2. Yoshizawa and Shiota alleged that the glutamate residues play an important role in the compensation of electronic charges after the formation of the copper–oxo species [32]. Both active species promote methane hydroxylation under physiological conditions. The reaction mechanism involves H-abstraction of methane by the oxygen atom in a mononuclear or dinuclear site, yielding a methyl intermediate that is stabilized by coordination to copper. The intermediate then undergoes methyl migration from the metal to the hydroxo group, giving rise to methanol. The transition energy of the C–H bond cleavage of methane and the formation energy of the methanol complex are computed to be 16.6 and −52.9 kcal mol^{-1} relative to the Cu^{III}–oxo species and methane, respectively, in monocopper-oxo species, and 17.6 and −49.2 kcal mol^{-1} relative to the bis(μ-oxo)$Cu^{II}Cu^{III}$ species and methane, respectively, in dicopper-oxo species. These mechanistic results using DFT calculations demonstrate that mono- or dicopper-oxo species as an active site has enough potential for the methane activation, and the complex with a dicopper site is a candidate for generating the active species for the methane activation.

According to the above knowledge, DFT was used to numerically assess the catalytic performance of a dicopper complex, which used H_2O_2 to catalyze the selective hydroxylation of benzene to phenol, for the hydroxylation of methane to methanol [33]. The focus was on the dicopper complex [Cu_2(μ-OH)(**6-hpa**)](ClO_4)$_3$ (**A**) with a dinucleating ligand 1,2-bis[2-[bis(2-pyridylmethyl)aminomethyl]-6-pyridyl]ethane (**6-hpa**), as shown in Fig. 10.

Complex **A** forms Cu^{II}O· and $Cu^{II}O_2$· species as active species, using H_2O_2 through three steps via intermediates with Cu_2O_2 and $(CuO_2H)_2$ cores [34]. DFT calculations showed that the Cu^{II}O· and $Cu^{II}O_2$· moieties in complex **1** are stably separate owing to the long Cu–Cu distance of 6.06 Å. The computed potential energy diagram of methane hydroxylation by complex **1** demonstrated that the reaction is initiated by the C–H bond cleavage of methane by the Cu^{II}O· core rather than $Cu^{II}O_2$· core, because the activation energies are 10.2 and 34.0 kcal mol^{-1}, respectively. The

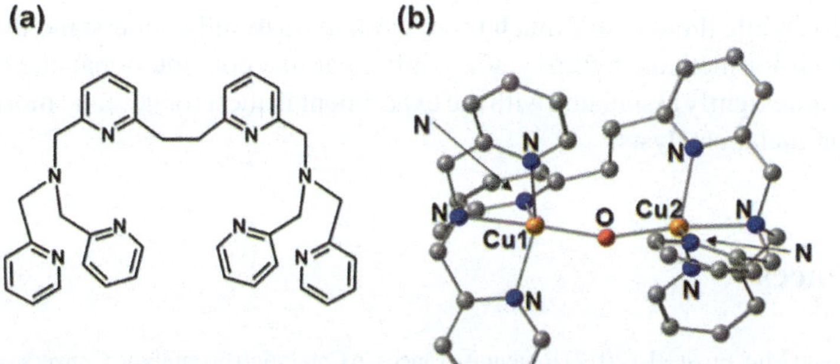

Fig. 10 **(a)** 1,2-bis[2-[bis(2-pyridylmethyl)aminomethyl]-6-pyridyl]ethane (**6-hpa**) and **(b)** the $[Cu_2(\mu\text{-OH})(\textbf{6-hpa})]^{3+}$ complex

H-atom abstraction forms methyl intermediate, in which the methyl group bonds to the copper center without the generation of methyl radical. The methyl intermediate then forms the methanol complex via an energy barrier of 10.8 kcal mol^{-1} in the C–O bond formation step, which is called the oxygen-rebound step. DFT calculations demonstrated that complex **1** contains a $Cu^{II}O\cdot$ species involved in the activation of methane C–H bonds, and the activation energy was shown to be lower than in the pMMO system, equaling 16.6 kcal mol^{-1}. Methane hydroxylation should take place in a two-step manner without radical species. Therefore, it is theoretically proposed here that complex **A** has a necessity precondition to hydroxylate methane to methanol under mild conditions.

6 Summary and Outlook

We discussed several examples of transition metal catalysts for organometallic and biomimetic hydroxylation of methane. Shilov, Periana, and Gilbert et al. adopted the organometallic approach and performed both experimental and theoretical investigations to demonstrate that methane C–H bond activation can be catalyzed by Pt(II) complexes in the presence of sulfuric acid. Moreover, they not only clarified the detailed reaction mechanism using DFT calculations but also proposed improved catalysts to design more efficient catalysts for methane hydroxylation. In addition, in the biomimetic approach, Sorokin et al. demonstrated that methane hydroxylation took place under mild conditions by using iron and copper complexes as sMMO and pMMO models. These experimental results inspired the theoretical study of Rajaraman et al., which elaborated the reaction mechanism as well as the electronic state of active species. Moreover, from the viewpoint of bio-inspired catalysts, Yoshizawa et al. showed that a dicopper complex can be available for an efficient catalyst for methane hydroxylation.

Finally, while there is still much more work to do to fully understand the reaction mechanism for methane hydroxylation, it is clear that computational chemistry will be even more tightly associated with the experimental quest for new and more efficient transition metal catalysts.

References

1. Adebajo MO, Frost RL (2012) Recent Advances in Catalytic/Biocatalytic Conversion of Greenhouse Methane and Carbon Dioxide to Methanol and Other Oxygenates. Liu G (ed) Greenhouse Gases - Capturing, Utilization and Reduction, IntechOpen, p 31–56.
2. Gunsalus NJ, Koppaka A, Park SH, Bischof SM, Hashiguchi BG, Periana RA (2017) Chem Rev 117:8521–8573.
3. Ravi M, Ranocchiari M, van Bokhoven JA (2017) Angew Chem Int Ed 56:16464–16483.
4. Holmen A (2009) Catal Today 142:2–8.
5. Benson SW (1965) J Chem Educ 42:502–518.
6. Hammes-Schiffer S (2017) Acc Chem Res 50:561–566.
7. Vogiatzis KD, Polynski MV, Kirkland JK, Townsend J, Hashemi A, Liu C, Pidko EA (2019) Chem Rev 119:2453–2523.
8. Shilov AE, Shteinman AA (1977) Coord Chem Rev 24:97–143.
9. Periana RA, Taube DJ, Evitt ER, Löffler DG, Wentrcek PR, Voss G, Masuda T (1993) Science 259:340–343.
10. Periana RA, Taube DJ, Gamble S, Taube H, Satoh T, Fujii H (1998) Science 280:560–564.
11. Alvarez-Galvan MC, Mota N, Ojeda M, Rojas S, Navarro RM, Fierro JLG (2011) Catal Today 171:15–23.
12. Hartwig JF (2010) Organotransition Metal Chemistry: From Bonding to Catalysis. University Science Books, Sausalito, Calif.
13. Balcells D, Clot E, Eisenstein O (2010) Chem Rev 110:749–823.
14. Siegbahn PEM, Crabtree RH (1996) J Am Chem Soc 118:4442–4450.
15. Shilov AE (1984) Activation of saturated hydrocarbons by transition metal complexes. D. Reidel Pub. Co.
16. Gilbert TM, Hristov I, Ziegler T (2001) Organometallics 20:1183–1189.
17. Ziatdinov VR, Oxgaard J, Mironov OA, Young KJ, Goddard WA, 3rd, Periana RA (2006) J Am Chem Soc 128:7404–7405.
18. Tsuji Y, Yoshizawa K (2018) J Phys Chem C 122:15359–15381.
19. Dalton H (2005) The Leeuwenhoek Lecture 2000 The natural and unnatural history of methane-oxidizing bacteria. Philosophical Transactions of the Royal Society B: Biological Sciences 360:1207–1222.
20. Hanson RS, Hanson TE (1996) Microbiol Rev 60:439–471
21. Kopp DA, Lippard SJ (2002) Curr Opin Chem Biol 6:568–576.
22. Balasubramanian R, Rosenzweig AC (2007) Acc Chem Res 40:573–580.
23. Himes RA, Karlin KD (2009) Proc Natl Acad Sci U S A 106:18877–18878.
24. Sorokin AB, Kudrik EV, Bouchu D (2008) Chem Commun:2562–2564.
25. Shilov AE, Shul'pin GB (1997) Chem Rev 97:2879–2932.
26. Periana RA, Mironov O, Taube D, Bhalla G, Jones CJ (2003) Science 301:814–818.
27. Bar-Nahum I, Khenkin AM, Neumann R (2004) J Am Chem Soc 126:10236–10237.
28. Periana RA, Bhalla G, Tenn WJ, Young KJH, Liu XY, Mironov O, Jones CJ, Ziatdinov VR (2004) J Mol Catal A: Chem 220:7–25.
29. Ansari M, Vyas N, Ansari A, Rajaraman G (2015) Dalton Trans 44:15232–15243.
30. Chan SI, Lu YJ, Nagababu P, Maji S, Hung MC, Lee MM, Hsu IJ, Minh PD, Lai JC, Ng KY, Ramalingam S, Yu SS, Chan MK (2013) Angew Chem Int Ed 52:3731–3735.

31. Yeh C-H, Yu SS-F, Chan SI, Jiang J-C (2018) ChemistrySelect 3:5113–5122.
32. Yoshizawa K, Shiota Y (2006) J Am Chem Soc 128:9873–9881.
33. Hori Y, Shiota Y, Tsuji T, Kodera M, Yoshizawa K (2018) Inorg Chem 57:8–11.
34. Tsuji T, Zaoputra AA, Hitomi Y, Mieda K, Ogura T, Shiota Y, Yoshizawa K, Sato H, Kodera M (2017) Angew Chem Int Ed 56:7779–7782

The manufacturer's authorised representative in the EU is Springer Nature Customer Service Centre GmbH, Europaplatz 3, 69115 Heidelberg, Germany. If you have any concerns regarding our products, please contact ProductSafety@springernature.com

Printed and bound by CPI Group (UK) Ltd, Croydon, CR0 4YY
30/09/2026
02225259-0013